Shut Subtopology Sieving in Construct-Steered Contrivance

Ploutos

Shut Subtopology Sieving in Construct-Steered Contrivance

First Edition August 2025

Written by Ploutos

TABLE OF CONTENTS

Page No's

LIST OF TABLES

LIST OF FIGURES

1. INTRODUCTION

1.1 OBJECT-ORIENTED PROGRAMMING

Object-Oriented Programming (OOP) is a programming language model organized around the real-time objects and data rather than actions that are required to be performed. Customer requirements vary from time to time, based on the business needs and growth options, because of this the software designed model should be extensible and reusable. In software paradigm, the architecture is evolved based on the software design principles and the main principle is to reuse of frequently used design patterns. Further, the domain-specific patterns that are best for a specific issue. A software product can contain many similar components at the design level. The unseen patterns of the programming system are important to know the system design and evaluate its quality. A better understanding of software design and architecture plays a vital role in the re-architecting process, which is an important aspect of maintenance task. Generally, recurring similar design structures are the most extended component of the design which can be good components to be utilized in the further proposal. It is very tedious and hard to identify. Identifying the frequent subgraphs from the given graph requires a thorough knowledge of graph theory and also a deep understanding of how to mine frequent subgraphs with respect to the domain. Frequent Subgraph Mining (FSM) delivers meaningfully structured information such as hot web access sequences, common protein structures, and Computational Molecular Biology. Frequent subgraph mining can also be used in fraud detection to catch similar fraud translation patterns from million
ns of electronic payments.

Object-oriented programming is the successor of procedural programming. Procedural programming describes implementation as groups of reusable code units (procedures) which define input and output parameters. Procedural programs consist of procedures, which invoke each other. The problem with procedural programming is that code reusability is hard and limited only procedures can be reused and it is hard to make them generic and flexible. There is no easy way to work with abstract data structures with different implementations.

The object-oriented approach relies on the paradigm that each and every program works with data that describes entities (objects) from real life. For instance, in the accounting software system that works with invoices, items, warehouses, availabilities, sale orders, etc are considered as objects. These objects describe characteristics (properties) and behavior (methods) of such real-life entities.

The main advantages and goals of OOP are to make complex software faster to develop and easier to maintain. OOP enables the easy reuse of code by applying simple and widely accepted rules.

1.1.1 Fundamental Principals

The Object-Oriented Programming (OOP) is constructed over four major principles: Encapsulation, Inheritance, Abstraction, and Polymorphism.

1. Encapsulation is the wrapping up of data and a member function which operate on the data into a single unit called the object. Free access to the data within an object is prevented by encapsulation. Access can only be done through the interface provided by the functions, such padding of data is mentioned to as data hiding.
2. Inheritance allows a class to acquire properties of the base class to derived class, inheritance supports reuse.
3. Abstraction allows us to do something very important features of the real-world objects in the program object. This process does not include the background details and explanations.

4. Polymorphism means one name, many forms. It conspicuous itself by having multiple methods all with the same name, but faintly different functions.

1.1.2 Object-Oriented Design

In designing, the object-oriented software can be termed as design patterns. Design pattern thoroughly names, give details and evaluates a significant and recurring design in Object-Oriented Design (OOD). The aim of the object-oriented design is to capture design issues and make object-oriented designs make more flexible, reusable, and elegant. They successful design helps developer for new designer with earlier experience. A developer who is well-known for such patterns can apply them right away to design issues without having to reinvent them.

The concept of design patterns was introduced by an architect named Christopher Alexander during 1970. Though he mentioned the concept with respect to architectural patterns of buildings, it's also applicable to object-oriented design patterns.

The most powerful work on design pattern was accomplished by Gamma, Helm, Johnson and Vissides universally known as Group of Four (GOF). According to GOF, for the creation of a reusable design, the design pattern names, abstracts as well as identifiers the important aspects of common design structures. The GOF design patterns can be described as communication among classes and objects which are customized for solving a design problem in a specified context (Gamma et al, 1995).

The Design patterns are organized into a form of a catalog. The design patterns with their names and intents are as follows

- The Abstract factory provides an interface to create families of objects (related/ dependent) without any specification of their concrete classes.
- Adaptor converts an interface of a class into another one according to the client's expectations and hence, overcomes the problem of incompatible interfaces thereby enabling the classes to work together.

- The bridge separates abstraction from its implementation to make them independent.
- Builder separates the construction of the complex object from its representation in order to create different representation with the same construction process.
- Chain of Responsibility enables handling of command objects by passing them to other objects by using the logic present in the processing of objects.
- Command encapsulates a request as an object and hence, enables to parameterized the different requests of the client such as log or queue request and also supports undoable operations
- Composite represents the object in a tree structure where each object represents the same interface. This enables clients to treat single objects and their formations uniformly.
- Decorator adds further functionality to a class at runtime. This enables flexibility to sub-classing for adding functionality
- Facade creates the unified interface of existing interface in the subsystem so as to handle common tasks easily.
- Factory Method focuses on object creation of specific type with specific implementation lets the subclasses decide as to which class to be instantiated.
- Flyweight performs sharing of common object properties by a large number of objects to save space.
- Interpreter deals with the implementation of specialized computer language that solves specific problems. The interpreter interprets sentences in language by representing the grammar of language along with an interpreter.
- Iterator enables sequential access of aggregate object elements by hiding their underlying representation.
- Mediator as the name implies, in a subsystem, it provides a centralized abstraction of interfaces. Mediator provides the loose coupling which enables objects to refer each other explicitly and also various objects interaction independently.

- Memento supports rollback mechanism by enabling the object to restore to its previous state without violating encapsulation by externalizing an object internal state.
- Observer defines a one-to many dependency between the objects so that an object changes its state, it raised an event which notifies othere objects and updates them automatically.
- Prototype Here, prototypical instance determines the type of object to be created by copying this prototype.
- Proxy offers an illusion by applying placeholder to another object in order to have control over that object.
- Singleton provides restrictions on instantiating a class to a single object and also makes it is globally accessible.
- State permits an alternation in the object behavior with alternation in its internal state i.e. allows objects type to change dynamically.
- Strategy is a collection of data structures and their selection based on the client
- Template Method describes the skeleton of a program, enables subclasses to define some steps of algorithms and also to redefine certain steps without affecting the structure of an algorithm.
- A visitor describes the separation of the algorithm from object and operations on the elements of an instance structure. It also enables to define new operations without modifying the elements classes on which it manages.

The design patterns in object-oriented/ software engineering have been classified by two criteria's

Purpose: These design patterns are again classified with their purpose into creational, structural, and behavioral.

Scope: It determines which patterns apply to classes and which patterns to objects and hence classifies them as class patterns and object patterns. The class patterns focus on the relation between classes and their subclasses which are accepted through inheritance at compile time whereas, object patterns focus on a relationship between objects which are more dynamic and changeable at runtime. The classified design patterns are shown in Table 1.1.

Table 1.1: Design Patterns classified with Purpose and Scope

<table>
<tr><td></td><td colspan="4">Purpose</td></tr>
<tr><td rowspan="14">Scope</td><td></td><td>Creational</td><td>Structural</td><td>Behavioral</td></tr>
<tr><td rowspan="2">Class</td><td rowspan="2">Factory Method</td><td rowspan="2">Adapter</td><td>Template Method</td></tr>
<tr><td>Interpreter</td></tr>
<tr><td rowspan="10">Object</td><td>Abstract Factory</td><td>Adapter</td><td rowspan="2">Chain of Responsibility</td></tr>
<tr><td>Builder</td><td>Bridge</td></tr>
<tr><td>Prototype</td><td>Composite</td><td>Command</td></tr>
<tr><td>Singleton</td><td>Decorator</td><td>Iterator</td></tr>
<tr><td rowspan="6"></td><td>Facade</td><td>Mediator</td></tr>
<tr><td>Flyweight</td><td>Memento</td></tr>
<tr><td>Proxy</td><td>Observer</td></tr>
<tr><td rowspan="3"></td><td>State</td></tr>
<tr><td>Strategy</td></tr>
<tr><td>Visitor</td></tr>
</table>

1.2 UNIFIED MODELING LANGUAGE

The Unified Modeling Language (UML) is an international industry standard graphical notation for relating software analysis and designs. When a uniform graphical notation is used, there is a little room of misinterpretation and uncertainty.

The UML represents a confederation of the theory and notations exhibited by the three amigos Grady Booch, Jim Rumbaugh, and Ivar Jacobson (Booch et al., 1999). The UML has been acknowledged as a standard by the Object Management Group (OMG). The OMG is a non-profit organization with about 700 members that sets standards for distributed object-oriented computation.

The objectives for UML to become a common language for creating models of object-oriented programming. The two significant parts of the UML are a Meta-Model and a Notation. The meta-model is an explanation of modeling language in UML. It clarifies the instances, attributes, and relationships required to represent the notions of UML within an application. The UML has included two important sections. There is a notation for modeling the static components of an outline, for example, classes, attributes, and relationships. The UML diagram includes use case diagram, class diagram, interactive diagrams, statechart diagram, activity diagram, component diagram and deployment diagrams. In this research work, class diagrams are considered and used for the analytical study.

Class Diagram is utilized in both the analysis and the design phases. During the analysis phase, a very high-level conceptual design is created. A class diagram might be created with the class names were possibly some pseudo-code-like phrases may be added to describe the tasks of the class. The class diagrams are developed during analysis phase for describing relationships between the classes in the problem domain, but implemented of the system does not refer. At the end of the design phase, how the system to be implemented should be developed is described by the class diagram. After the completion of the design phase, the class diagram was created. The class diagram provides implementation information which including the class names, the operation and member functions of the classes, and the associations among classes.

The class diagram depicts the type of objects in a programming system and the different kinds of static associations that exist among them (Bruegge and Dutoit, 2000). In UML a class diagram consists of three rectangular horizontal chambers. The upper chamber clenches the name of the class, middle chamber clenches the catalog of attributes or data members, and the last chamber clenches the catalog of operations. The whole UML graphical notation of a class is visualized in Figure 1. 1

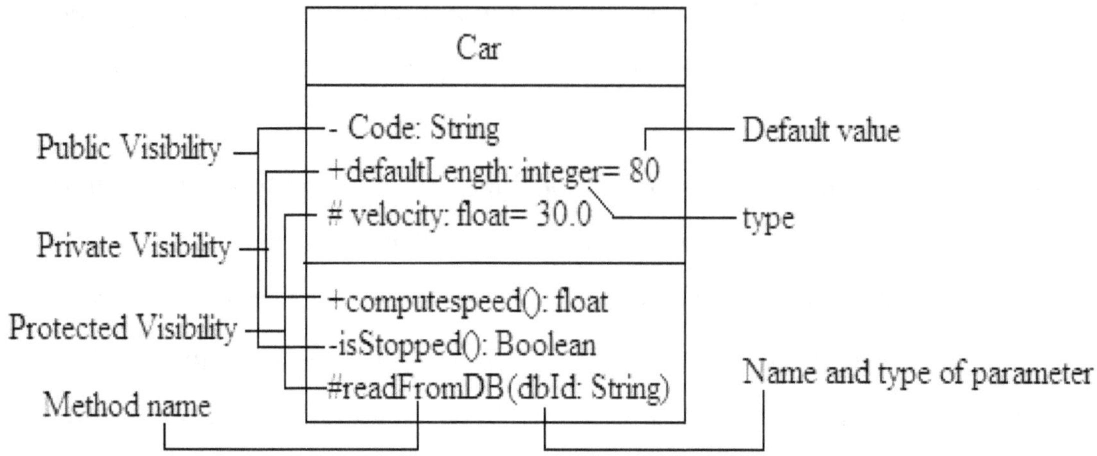

Figure 1.1: UML Notation for Class Name, Attributes/Member Functions and Operations/Methods

The different relationships between the two instances of the class are shown in Table 1.2 with associations having a multiplicity that signifies how many instances of each class can logically be involved in known relationships. Table 1.3 shows the multiplicity notations.

Table 1.2: Relationship Names and Their Symbols

S.no	Relationships name	Description	Symbol
1	Association	It specifies that class uses an attributes of another class. In case there is line without an arrow, then both classes hold information about other class.	 
2	Generalization	An empty arrow with the triangle at one side, pointing from the subclass to the superclass. The subclass inherits all the attributes and methods of the superclass and may inherit override method.	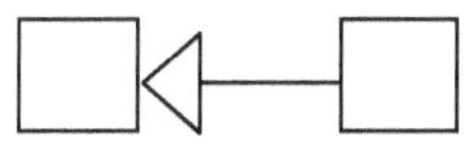
3.	Aggregation	It communicates the information that one instance is part of another instance, but independent of a lifetime.	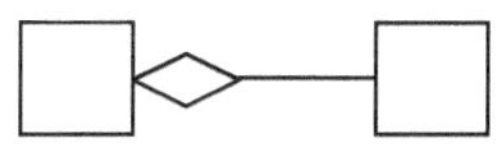
4	Composite	It is a strong form of aggregation	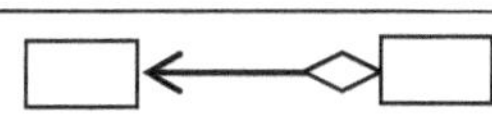

Table 1.3: Cardinality and Modality

Cardinality and Modality	UML Symbols
One- to- one and mandatory (one and only one)	1
One-to-one and optional (zero or one)	0..1
One-to-many and mandatory	1..*
One-to-many and optional	*
With lower bound l and upper bound u	l….u
With lower bound l and no upper bound	l….*

1.3 SOFTWARE REUSE AND RESTRUCTURING

The high costs of developing software motivate the reuse and evolution of existing software. Software reuse in its broadest sense involves reapplying knowledge about a software system, to reduce the efforts of developing and maintaining another system. Closely related to software reuse is software maintenance, where knowledge about a software system is used to develop a version that refines or extends it.

Approaches that support reuse address one or more of the following four important aspects (Biggerstaff and Richter, 1989):

1. Obtaining frequently reused component. It is tedious to find an exact match from the derived graph, quite involves similar component.
2. Analyzing and identifying the component. Analyzing the behavior of a component is utmost important to reuse the component. It is difficult to build a mental model from the reused component (Brooks, 1987; Parnas, 1979).
3. Updating a component or a collection of components. It is human intensive to identify the behavioral changes for a component such as what changes are needed and identifying depended objects.
4. Originating the components together. The construction procedure can be tedious, particularly when the component is being used as multipurpose component such as to construct other composite structures or as independent entity.

Refactoring of Software architecture and design relates to the above mentioned four important aspects of reusability. Redesigning software design can make it simpler to understand for improving the code readability and can help in identifying the association, inheritances, coupling, and cohesion of components. Modifying the behavior and structure of a component to make it more reusable, these parts can be simpler to construct a software application.

Enhancing the programming design requires, sometimes requests to be redesigning prior to reuse. Arnold (1986) defines software redesigning as the change of software for (1) better understanding and improved code readability or (2) less vulnerable to the design issue and defects when enhancement to the software are required.

A major goal of software restructuring is to preserve or increase the value of a piece of software. Restructuring a software with enhanced programming techniques for improving the performance, scalability, and usability of the software to the existing system. Software restructuring is most offen used during software maintenance, wherever the software developer has a lack of knowledge in existing software behavior. Illustration of refactoring of design structures in Figure 1.2.

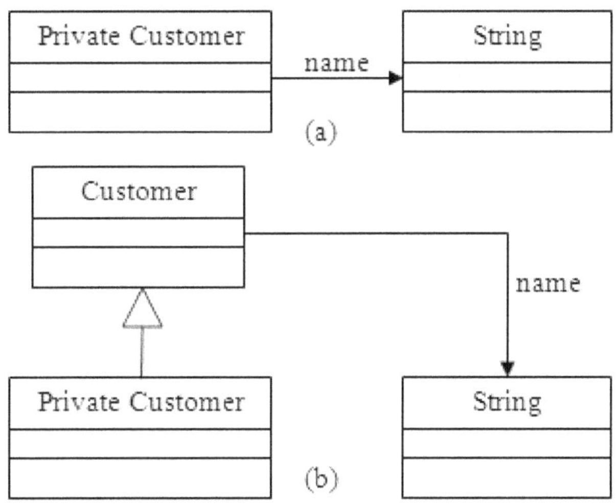

Figure 1.2: a) Before Refactoring b) After Refactoring an Exemplary Object–Oriented Model

1.4 DATA PREPROCESSING

Data preprocessing is a data mining procedure that occupies converting a raw data into a logical format. A real database is habitually an incomplete, inconsistent and noisy data due to their typically gigantic size. Data preprocessing is an established technique for resolving such issues. Data preprocessing refine the raw data for supplementary processing.

Data preprocessing consist of:

1. Data cleaning method fills in missing values, smoothness by identifying outlier noise, and inconsistencies in the data to be corrected. This technique provides a way of handling missing values and smoothes out the noise in a variable using binning, regression and clustering technique.
2. Data integration method consolidates data from numerous sources to shape a coherent data store. The contributions towards the smooth data integration are Metadata, correlation analysis, data conflict detection, and the resolution of semantic heterogeneity.
3. Data transformation schedules change over the data into proper structures for mining. For example, attribute data may be normalized so as to fall between a small range, such as 0:0 to 1:0.
4. Data reduction is utilized to reduce the presentation of the data while reducing the loss of content. Their techniques such as attribute subset selection, data cube aggregation, numerosity reduction, dimensionality reduction, and discretization.

1.4.1 Data Transformation

In data transformation, the data are modified into forms more appropriate for mining. Several techniques may be viewed as data transformation techniques:

1. Smoothing techniques are binning, clustering, and regression which are established to eliminate noise from the data.
2. Generalization is low-level descriptions that are substituted by higher-level descriptions.
3. Normalization is recalling of attributes
4. Attribute creation is generating new attributes and supplementary to the existing ones.

Data transformation such as normalization may enhance the accuracy and efficiency of data mining algorithms involving neural networks, nearest neighbor, and clustering classifiers. Such schemes offer to improve results if the data to be evaluated have been normalized, that is, scaled to specific ranges such as [0.0 to1.0] (Jiawei and Kamber, 2001).

An attribute of a data set is normalized by scaling its value so that they fall inside a tiny-specified range, such as 0.0 to 1.0. Normalization is particularly useful for clustering and classification algorithms. There are numerous schemes for data normalization that include min-max normalization, z-score normalization, and normalization by decimal scaling.

Min-max normalization presents a linear transformation on the original data. Min-max normalization technique maps the original value A to the normalized value A' in the pre-defined boundary [new_max_C, new_min_D]. The min-max normalization is premeditated by the following principle:

$$A' = \frac{[\text{A}-\text{minimum value of A}]}{[\text{maximum value of A}-\text{minimum value of A}} * [\text{new\_max}_C - \text{new\_min}_D] + \text{new\_min}_D \quad 1.1$$

Z-score Normalization is also known as zero-mean normalization. In Z-score normalization technique, the values for an attribute 'A' are normalized depends on the mean and standard deviation of attributes. A value of A is normalized to A' by the following principle:

$$A' = \frac{\text{A}-\text{mean(A)}}{\text{std(A)}} \quad 1.2$$

Normalization by decimal scaling normalizes by moving the decimal point of the value of attribute A. The number of decimal points moved depends on the maximum absolute value of A. A value of A is normalized to A′ by the following formula:

$$A' = \frac{A}{10^m} \quad 1.3$$

Where m is the smallest integer such that Max (|A' |) < 1.

1.5 GRAPH PRELIMINARIES:

This section provides an overview of basic concepts in graph theory and also the notations used in the rest of the thesis.

Graph- A graph represents nodes (vertices) and edges and denoted by G. G is a set of (V, E) embodying of a collection of nodes V and a set of links or connections $E \subseteq V \times V$, that is every link $e \in E$ represents a set of nodes (v_1, v_2).

Undirected Graph- For each link $(v_1, v_2) \in E$ of graph G = (V, E), there exist $(v_2, v_1) \in E$, graph is said to be undirected and the links are known as edges 'E' of Graph G. If a link is uniformly represented by $(v_1, v_2) = (v_2, v_1) = e \in E$ with v_1 and v_2 in V.

Simple Graph- If the undirected graph has cyclic links nor multiple links or connections then it is known as Simple Graph.

Regular Graph- If the undirected graph has the same number of neighbors for all the vertices in Graph G.

Weighted Graph- A graph G = (V, E) is connected with a real value and is said to be weighted w (e) and i.e., associated with each edge e and w (e) is known as the mass of the edge e. A set of vertices $(v_1, v_2) \in V^2$, vertices (v_1, v_2) does not belong to edge E, if the weight of the vertices (v_1, v_2) is considered to have a zero i.e., $w(v_1, v_2) = 0$.

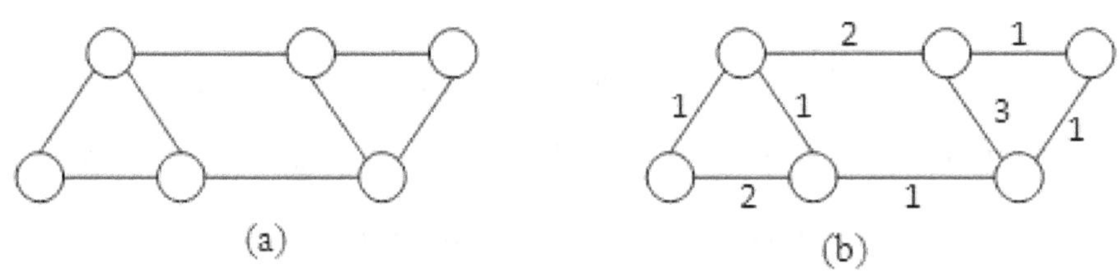

Figure 1.3: (a) Graph G (V, E) with 6 nodes and 8 links. (b) Weighted graph with weights on links

Adjacency Matrix- For a simple graph G= (V, E), the Adjacency matrix (Adj) has n × n dimension, such that ∀ (i, j) ϵ $\{1 \ldots n\}^2$

$$(\mathrm{Adj})_{ij} = \begin{cases} 0 & if\ i = j \\ \mathrm{w(i,j)} & else \end{cases} \quad 1.4$$

is called the adjacency matrix of the graph G.

Degree Matrix- For a simple graph G= (V, E), the matrix Degree (Deg) has n × n dimensions, such that ∀ (i, j) ϵ $\{1 \ldots n\}^2$

$$(\mathrm{Deg})_{ij} = \begin{cases} \mathrm{deg\ (i)} \sum_{k=1}^{nv} \mathrm{w(i,k)} & if\ i = j \\ 0 & else \end{cases} \quad 1.5$$

is called the degree matrix of G.

Graph Partition- let V be a regular set. A set P of a subset of V is called a Partition of V if:

- P is a non empty set of elements.
- The element of P is pairwise dislocate
- The combination of elements of P is the same to V.

Graph Isomorphism- Two graphs $G_1 = (V_1, E_1)$ and $G_2 = (V_2, E_2)$ are isomorphic if they are topologically matching to each other; that is, there is a mapping from G_1 to G_2 such that every link in E1 is mapped to a single edge in E2 and vice versa. In the case of labeled graphs, this mapping has to defend the labels on the vertices and edges.

Subgraph- A graph $G_2 = (V_2, E_2)$ is a subgraph of another graph $G_1 = (V_1, E_1)$ iff $V_2 \subseteq V_1$, $E_2 \subseteq E_1$ ^ $((v_1, v_2) \in E_2 \rightarrow v_1 \in V_2$ and $v_2 \in V_2)$.

Subgraph Isomorphism- Given two graphs $G_1 = (V_1, E_1)$ and $G_2 = (V_2, E_2)$, the difficulty of subgraph isomorphism is to discover an isomorphism between G_2 and a subgraph of G_1, that is to conclude or not G_2 is included in G_1.

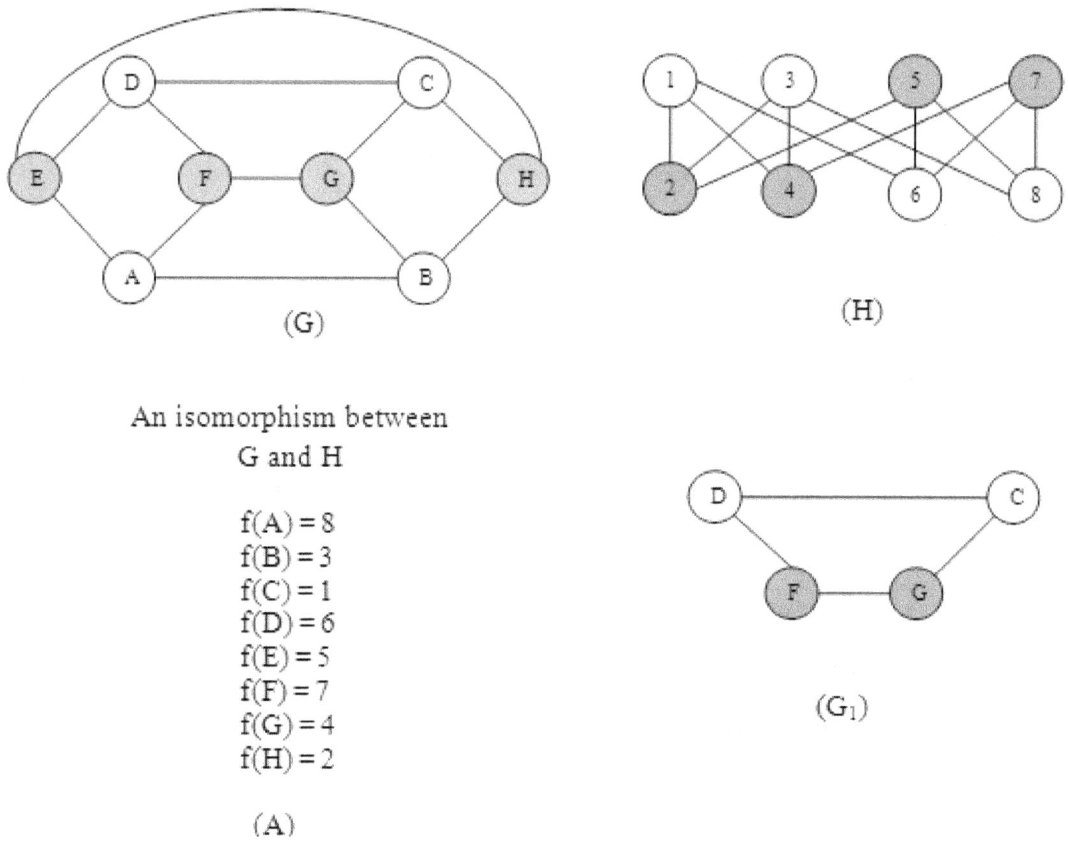

Figure 1.4: Graph G and H are Two Different Graphs which are Isomorphic by (A), G1 is the Subgraph of the Graph G

Frequent Subgraph- A frequent subgraph is a graph whose support is not less than a minimum user-specified support threshold. $G_D = \{G_1, G_2, G_3 \ldots G_m\}$, support or frequency of a subgraph g is the number of graphs in G_D where g is a subgraph.

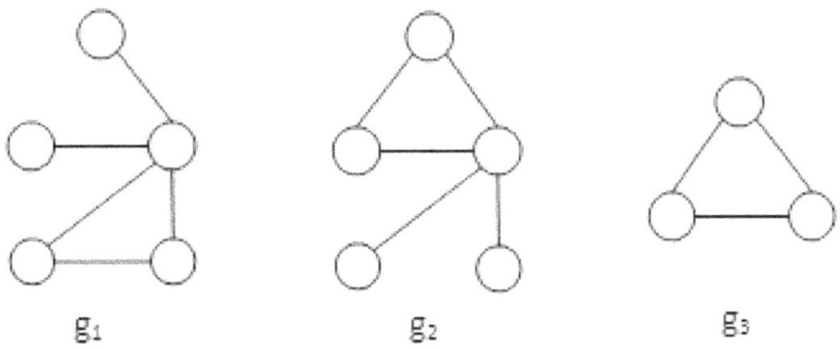

Figure 1.5: g_1 and g_2 are Two Graphs in Database and g_3 is Frequent Subgraph in a Graph Database.

1.6 GRAPH PARTITIONING

The term partitioning itself defines the achievement of generating a separation by dividing a set into a number of parts. A large number of engineering problems are solved by Partitioning. Many applications for partitioning are data mining, designing of VLSI, parallel computing, network load balancing, image processing and air traffic controls etc.

Generating a separation consists of dispensing a set of instances into various subsets which may be useful to compare them. Consequently, each object is linked with some other objects. When each object is linked with the entire other objects, then the combination is known as clustering. If the set of all these objects is finite, then the entire environment would be met to transpire the links between things. The subsequent step will produce a partition of this graph.

Partitioning a set of instances commonly tries to dispensing these objects into components having tough internal connections and fragile connections between components. This objective of partitioning diverges to the specific problem which is to be solved.

A graph can be partitioned by two common approaches include edge-cut and vertex cut. These approaches have partitioned the graph into the required number of disconnected parts.

Edge-cut partitioning- There are several edge-cut based algorithms for partitioning a graph, consists of spectral graph partitioning and minimum edge-cut based algorithms.

Cut function-A graph G = (V, E). Given V_x and V_y are two subsets of vertex V, and classify the cut between these two subsets as:

$$\mathrm{cut}(V_x, V_y) \sum_{u \in V_x} \sum_{v \in V_y} w(u, v) \qquad 1.6$$

The study, given $P_k = \{V_1, V_2, \ldots, V_k\}$be a partition of V into k parts.

The Cut function is used by Kernighan and Lin (1970). It intends to minimize the sum of the mass of the edges among divisions of the partition P_k.

$$\begin{aligned} \mathrm{cut}(P_k) &= \sum_{i<j} \mathrm{cut}(V_x, V_y) \\ &= \frac{1}{2} \sum_{i=1}^{k} \mathrm{cut}(V_x, V - V_x) \end{aligned} \qquad 1.7$$

Ratio cut function is introduced by Wei and Cheng (1989). It targets to minimize the ratio among its cut and its weight of the edges for each part.

$$\mathrm{ratio}\ (P_k) = \sum_{x=1}^{k} \frac{\mathrm{cut}(V_x, V - V_x)}{w(V_x)} \qquad 1.8$$

The function broadly utilized in graph partitioning is known as Normalized cut (Shi and Malik, 2000). An image segmentation issue was replaced by ratio cut. It targets to minimize, for each component, the ratio between the sum of the mass (weight) of the edges neighboring to one of its node and the sum of the mass of the edges neighboring to at least one of its nodes;

$$\mathrm{norm}(P_k) = \sum_{x=1}^{k} \frac{\mathrm{cut}(V_x, V - V_x)}{\mathrm{cut}(V_x, V)}$$
$$= \sum_{x=1}^{k} 1 - \frac{\mathrm{cut}(V_x - V_x)}{\mathrm{cut}(V_x, V)} \qquad 1.9$$

A familiar property of all the partitioning method is based on selected pair of edges to be cut from a graph to resulting graph is separated into various connected parts. As shown in figure 6.a

Vertex-cut partitioning: While edge-cuts result in partitions that are vertex-disjoint, in vertex-cuts the data vertices can serve as bridges between the resulting graph data separations in Figure 6.b; the cut passes through the vertices of the graph.

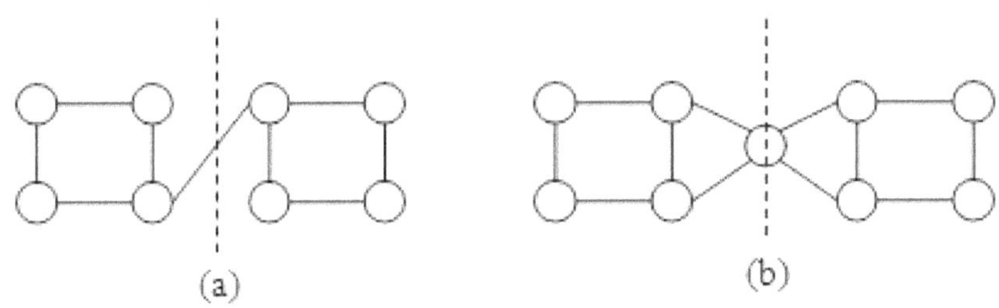

Figure 1.6: a) Edge Cut b) Vertex Cut

1.7 GRAPH CLUSTERING

Any non-uniform data include basic structure suitable to the heterogeneity of the graph data. The procedure of recognizing this structure in terms of combining the data elements is known as clustering, also called classification of graph data (Kleinberg and Tardos, 2002) The out coming groups is known as clusters. The grouping habitually depends on some similarity measure distinct for the data objects. Clustering is very much related to unsupervised learning in sequence identification systems (Duda et al., 2001). An assignment in unsupervised learning is to organize a data set into two or more classes based on a similarity measure over the data, without return to any a priori information on how the classification should be done.

Graphs are produced by a set of vertices and a set of links that are connections between two nodes. Graph clustering is the task of grouping the nodes of a graph into clusters by considering the associated structure of the graph, such that there are sufficient links within each cluster and relatively few between the clusters. Graph clustering in the logic of clustering the nodes of an input graph into clusters, while framing clusters based on structural similarity should not be confused. The measure of graph similarity is addressed in other literature (Luo et al, 2002; 2003; Robles-Kelly and Hancock, 2005) even though many of these methods are closely related to the job of discovery clusters within a graph.

1.8 GRAPH MINING

Graph mining has a huge importance to generate practical applications because of the structured and the data widely arise in different real-time fields including biological, material science, social networks, protein network, chemistry, bioinformatics and fraud detection. The primary goal of graph mining is to identify the possible sequences, mining topological structures in graphical data.

The finding of frequent subgraphs generally consists of two phases. In phase one, it produces frequent subgraph candidates while the occurrence of each candidate is validated in the second phase. Most of the studies in frequent subgraphs finding focuses on the first phase since the next step involve subgraph isomorphism that is NP-complete.

Some of the approaches which identify candidate subgraphs are scheduled below:

1. Level-wise joins: Two subgraphs (g_1 and g_2) of size 'k' are combined together to form a (k+1) candidate subgraph.
2. Rightmost path extension: vertices are added on the rightmost path of a k-subtree to form a (k+1) subtree.

   And other approaches such as Extension and Join, Right- and- left the tree to join and Equivalence class based extension are also used in candidate generation phase.

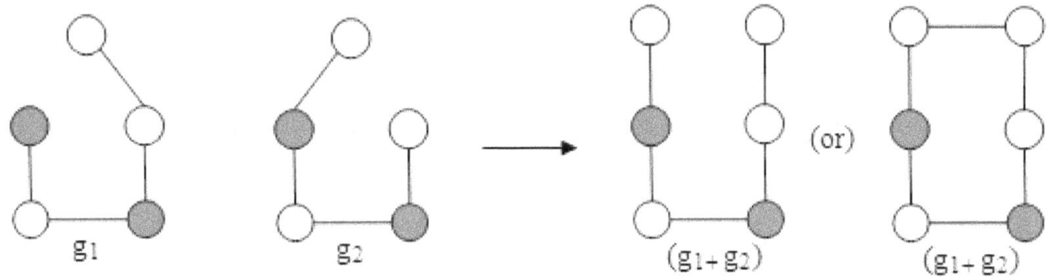

Figure 1.7: Two subgraphs of size k are merged together to form candidate subgraphs of size k+1

In data mining widely used methods graph mining methods such as partial methods also shares identical measurements those are information entropy, information gain, Minimum Description Length (MDL), gini-index and others (Holder et al., 1994; Yoshida et al., 1994). Among all of them well-known approaches, a mechanism evaluated to evaluate the support of each item from the basket analysis (Agrawal and Srikant, 1994). Its basic scheme can be represented as follows:

Given a graph data set $G_D = \{G_1, G_2, G_3, \dots \dots, G_n\}$

$$\sup(G_s) = \frac{|D_g|}{|G_D|} \quad 1.10$$

D_g represents the number of graphs in G_D where g is a subgraph; sup (G_s) is the supporting dataset of g; $|D_g|$ is complete support of g; and $|D_g|$ is the total number of graphs in G_D. A subgraph g can be described as a recurring subgraph if and only if its support is no less than a least supported threshold, δ by experimented its occurrence sup (G_s)= | D_g | / |G_D| ≥ δ (Yan, 2003).

There are two basic approaches to the frequent subgraphs mining problem. (1) Apriori-based approach and (2) Pattern-growth approach.
Apriori-based frequent subgraph mining algorithms contribute similar characteristics to the apriori-based recurrent itemset mining algorithm. The search for repeated graphs initiates with graphs of tiny size and continues in a bottom-up manner. At every call, the size of newly identified repeated subgraphs is increased one by one. These two subgraphs are generated by joining two parallel but slightly dissimilar repeated subgraphs that were exposed already. The regularity of newly formed graphs is verified. Two typical apriori-based frequent mining schemes have already been exposed AGM algorithm was suggested Inokuchi et al (2000) and FSG was introduced by Kuramochi and Karypis (2001).

Unlike apriori-based mechanism, the algorithm pattern-growth mining extends a recurring graph by accumulating a new edge in all the probable positions. An important issue with the edge extension is the similar graph can be investigated several times. To solve these problems, Yan et.al (2000) proposed a rightmost extension technique in gSpan, where the extension takes place on the rightmost path. A rightmost path is a straight path from the opening vertex v_0 to the final vertex i_n, according to the depth-first search (DFS) on the graph.

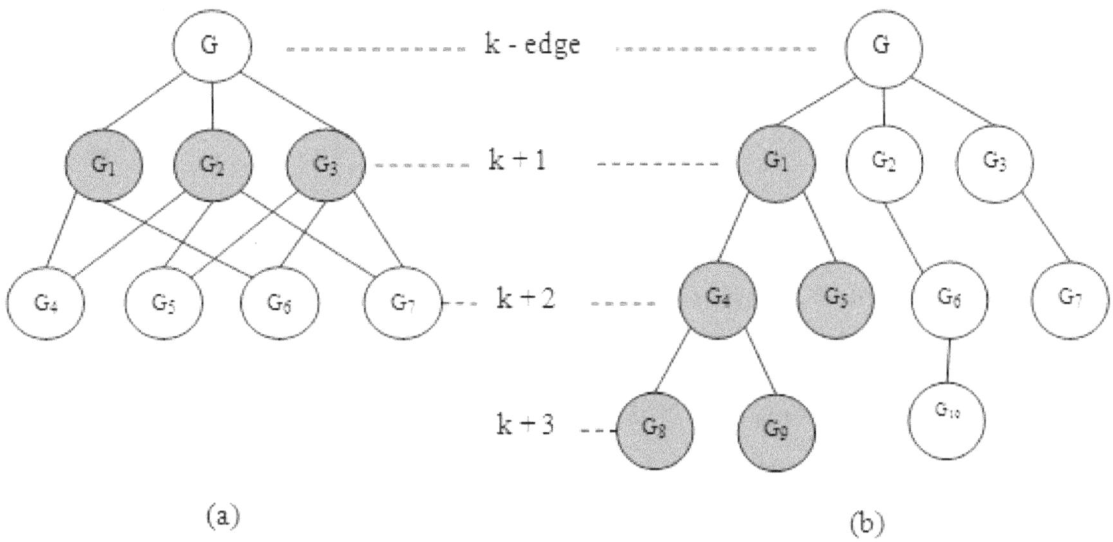

Figure 1.8: (a) Apriori Based Approach (b) Pattern Growth Approach

1.9 MOTIVATION FOR RESEARCH WORK

Discovering an identical frequent design structure in object-oriented design achieves a repeated structures in a software or Object-oriented design. The obtained structures can be helpful for developers to develop a new design by reuse the structure or refactor the structure. The large and complex software systems are difficult to understand by developers or designers. So that large system have to partition into multiple sets by graph partitioning methods and finding the frequent structures by mining technique.

1.10 OBJECTIVES OF RESEARCH WORK

The main objectives of the proposed work are:

1. To generate a software model graph for a source code.
2. To propose a methodology for partitioning a complex software model graph into modular sub-graphs.

3. To suggest a close graph mining algorithm for detecting frequent design structures from the partitions.

1.11 SIGNIFICANCE OF THE PROBLEM

Object-Oriented Design (OOD), a software paradigm to solve a problem that can be re-used in many different applications. To design a new application, existing frameworks or design patterns can be reused. Further understanding a software design is not an easy assignment for a developer. Consequently, it becomes necessary to identify the similar design structures so that multiple cross-functional teams can understand each of the component/modules of the system, this will help to refactor and get a better quality of the software effectively by different teams.

To understand legacy applications, developers should have in-depth knowledge of the application and domain specific. And also the developers have to manually identify which designs principles are reused thoroughly/repeatedly. It is difficult to understand the functionalities of large object-oriented systems because of their topological and behavioral complexity. It is even more difficult to extend or modify them. Thus, it becomes necessary to divide large and complex systems into smaller parts.

1.12 THESIS ORGANIZATION

Chapter I deals with the introduction of various important concepts related to object-oriented programming, design patterns, UML class notation, graph theory definitions, graph partitioning techniques, clustering techniques for graph data and various approaches for mining the graph data to identify the frequent structures and patterns are discussed in brief.

Chapter II discusses the recent developments in various topics related to the proposed system, like the creation of software model graph for source code, Graph partitioning/graph clustering techniques to divide the large graph into multiple parts, and Graph mining methods on ILP, apriori, and pattern-growth for a multiple graph data.

Chapter III presents the main contribution of the work. The methodology for the development of the proposed system and its architecture is illustrated in this chapter. Various phases of the development process are explained in three phases namely, generating of class-oriented model graph phase, graph partitioning phase, and frequent subgraph mining phase.

Chapter IV described the proposed approach of the first phase, which is named as a class-oriented model graph. In chapter cover syntax tree, semantic tree, type systems and finally construction of class-oriented model graph.

Chapter V describes the proposed algorithm of the second phase, which is named as an unfolding and boosting based graph partitioning algorithm. It is a repetitive process of unfolding and boosting operations and illustration with an example.

Chapter VI describes proposed an algorithm of the third phase in methodology. The proposed algorithm is used to discovering the identical frequent subgraph in object-oriented models with illustration.

Chapter VII detailed experimental study of generating a class-oriented model graph, the experimental study of the proposed UB-GPA algorithm and the experimental study of generating close frequent subgraph mining algorithm. The results of experimental studies with hypothetical and real-world datasets are discussed.

Chapter VIII concludes the research work and suggests further extensions to this research and also includes references and list of research publications out of this work.

2. LITERATURE REVIEW

The review of literature given in this chapter is centered upon new studies for the construction of source code to model graph, abstract syntax tree (AST), software reusable, restructuring, graph clustering, graph partitioning and frequent subgraph mining approach based on graph mining approaches for data mining applications.

2.1 REPRESENTATION OF SOURCE CODE INTO GRAPH

Kleppe and Rensink (2008) proposed a proper extension of type model graphs with notations that are common in UML. The authors suggested unambiguous, formal semantics to the UML using the theory of graphs and graph transformation. Generally, class diagram can be treated as complex type graph. Class diagram structures are much richer, which is incorporated in their framework by using constraints of the graph. The constraints which are used to make possible to change in the given semantics to include or exclude certain semantic elements. In addition, not only the clarity provides semantics for both diagram types but relations between them. The outcome should be reviewed not on originality, although on completeness, adaptability, and usability. If there is no well-defined forum where this type of effort can receive recognition, there will be no incentive and the gap between theory and practice may remain with forever.

Rajani and Roopa (2013) demonstrated a source code to module dependency graph using class dependency analyzer, which is used to recover the high-level structure of a software system. For code analysis, they used AT&T's CIA tool for C (Chen, 1995) and Acacia for C++ (Chen et al., 1997) and Chava or java (Korn et al, 1999). All the module-level dependencies can be stored in a database, executed an abstract window toolkit script to a database query, sort the query outcome and producing a textual representation of the graph.

Neamtiu et al., (2005) presented an approach which depends on partial syntax tree matching and can track easy modification to global variables, types, and functions. These modifications can exemplify aspects of software evaluation which helps to answer advanced level questions. They commenced by parsing the two implementation versions to produce syntax trees, which are traversing in equivalent to collect type and name mappings. With the mappings at hand, changes are detected and reported to the user, either directly or in the summary form. The syntax matching algorithm illustrates how modifications are detected and reported, describes the development and its performance. It is described that the tool was efficient and also some insight into the evaluation of software.

Baxter et al., (1998) presented an approach for detection of clone using syntax trees. Duplicate detection work was partial to the finding of accurate textual matches. Practical methods are used for detecting exact and near miss clones for arbitrary fragments of program source code. Since recognition of the program structure copies can be factored out of the source using conventional transactional schemes. The direct method of parsing technology is implemented to detect duplicates in arbitrary language creates and computer macros that agree to the removal of the duplicates without disturbing the operation of the curriculum. The method is applied to the real application of reasonable scales and confirmed previous estimates of clone density of 7-15%. Automated methods can find and eliminate such duplicates, decreasing the value of this constant and cost of the maintenance. Duplicate detection tools also have better potential for aiding domain analysis.

Jeffrey L et al., (2009) proposed a six grammar annotations - omission, labeling, Boolean access, list formation, inlining, and superclass development that permit syntax for a language to be characterized depends on its tangible syntax. An annotated grammar, produce both a parser and a rewritable abstract syntax tree. Concretizing syntax tree permits it to protect the designing of the creative code even after rewriting. Moreover, a syntax tree generator based on the method can couple the produced AST-builder with a pseudo-preprocessor to permit describes and control of preprocessed code. In Ludwig, syntax tree generator had implemented and had used to generate the rewritable AST in a restructuring the annotated grammar was more than a request of extent smaller than the generated code, and the overhead of concretizing ASTs was very reasonable.

Singh et al., (2013) proposed a software quality tool for measuring the different code metrics for C# source code using syntax tree. Nfactory libraries are used to generating abstract syntax tree of the implementation.

Virdi and Singh (2012) proposed different types of coupling i.e. static and dynamic coupling. These metrics performed under the different environments and calculate the mean and the standard. The value of standard deviation is useful in judging the representativeness of the mean and quality of software system.

2.2 REUSE AND REFACTORING OBJECT-ORIENTED DESIGN

An Object-Oriented system can be reused or expanding simply by grouping the components of presented classes in various ways. Often, object-oriented system can be reused after restructuring the component. There are various causes for this (Casais, 1991; Johnson and Foote, 1988)

1. When software application was developing, it faces hard to determine in what classes represent the essential concepts for a particular application and how they are interrelated.
2. Even after a system has matured through several iterations, sweeping structural changes might still be necessary. The software system must operate in an environment that is constantly changing, and the software system must satisfy user needs that are constantly changing as well.

3. When applying reuse component in projects it may raise new issues. So that system should be partitioned differently, suitable to organizational and other aspects, in order to reuse by a new project. Accordingly, some refactoring can be needed to effect reuse.
4. An object-oriented system can concurrently supply several functions in sub-classing, like code sharing, type validation, and modeling generalization and specialization relations. It is tricky to reconcile all of these ideas are achieved when required restructuring is proposed.

When the object-oriented scheme is introduced means this issue does not disappear. One of the fundamental difficulties in implementing software is that the software must be conventional to human institutions and systems that interface with it, and those forces are persistently changing (Brooks, 1987).

Lee et al., (2001) presented a literature about restructuring process in object-oriented, which recommend that the developers can reuse instances until to defeat the limitation of code reuse. Although, developers are having the positive outcomes such as the performance of the system, optimizing structures and checking the merits and demerits of this process in the D2D system of D Company and computed the result.

Schulz et al., (2009) describing object-oriented system schema level with classes, relations, methods, and inheritance with the instance level with objects, relationships, operations, and messages. Homeomorphisms' are utilized for the instance level types with a description of restructuring which specify the addition, folding, and unfolding of schema elements. a categorical framework is presented which permits deriving object migrations from representation in such a way that implementation and procedures to the old representation are appropriately migrated into programs and procedures to the new representation.

2.3 GRAPH CLUSTERING

Dongen (2000) suggested a Markov Cluster (MCL) Algorithm, which converts the values of a transition matrix towards either 0 or 1 at every progression in a random walk until the problematic conditions are pleased. When the Hadamard power for every transition probability value is divided by the summation of each column, the rescaling procedure defers a transition matrix for the further stage. After repeating alternatively for around 20 times between two stages random walk and probability change, the procedure will, at last, achieve a steady state in which the whole graph is partitioned into a set of 'hard' clusters. Analysis of MCL is O (N3), where N is the number of nodes. MCL is compared with RRW and navie nearest neighbor approach were on a biological protein network for yeast, as well as a noisy version of the network. Proteins with the same biological function should cluster together. The resulting clusters were compared to know protein groupings.

Xanthos (2006) proposed a spectral graph partitioning method to decompose the object-oriented software system. The methodology depends on an iterative process for separating the graph in order to classify dense communities of classes thereby minimizing the message between the modules of the system. This is realizing because it finds the minimum cut set of edges in each call. Therefore, it can also be utilized to identify the modules that should be assigned to diverse machines, in a distributed environment.

Chatzigeoriou et al., (2006) presented a graph theory application for object-oriented software engineering. Four different graph theory applications have been established, regarding detection of 'God' classes, clusters, identification of sequences and scale-freeness of an object-oriented system. Methodology for identification of "God" classes is illustrated with examples (Chatzigeoriou et al., 2004; Riel 1996). Spectral graph partitioning algorithms for placing strongly connected cluster classes. Blondel et al., (2004) suggested an algorithm, which is based on Kleinberg's analysis is utilized for the recognition of design patterns and the scale-free phenomenon that exist purely in network technologies such as class association graphs and graph objects in object-oriented systems. Progression of scale-freeness for numerous versions of three different open-source systems for object-objected system tends to become more or less scale-free.

Czibula et al., (2007) presented a new clustering algorithm based on hierarchical agglomerative utilized for object-oriented systems restructuring. This methodology wants at classifying a partition of a software component that corresponds to an enhanced structure of it. Hierarchical Agglomerative clustering algorithm for Restructuring Software systems (HARS) can be utilized in the combining step of Clustering Approach for Refactoring Determination (CARD) in order to re-group classes from the system. This methodology can be helpful for supporting software engineers in their regular works of restructuring software components. The open source JHotDraw is a java based GUI framework for mechanical and structured graphics is evaluated highlighting its merits in comparison with presented approaches.

Erdemir et al., (2011) proposed an object-oriented software clustering approach based on the adoption of the fast community detection algorithm. It is an agglomerative hierarchical clustering algorithm specifically designed for discovering neighborhoods in social network graphs within the small-world behavior. The algorithm was implemented on the directed weighted graph. Authors evaluated the algorithm and compared with previously designed approaches. It is observed that the algorithm has considerable speed average over other algorithms.

Zhang et al., (2012) proposed a simple but successful graph-based agglomerative algorithm for high-dimensional clustering data. They explored two different roles in graph theory, in-degree, and out-degree, in the framework of clustering. The average in-degree replicates the density near a sample, and the average out-degree differentiates the local geometry around an example. Based on such approaching, they have defined an affinity measure of clusters via the product of average in-degree and average out-degree. The product-based affinity make algorithm robust to noise. The three main advantages of algorithm are: performance, trouble-free implementation, and high computational efficiency. They tested the algorithm on two different computer vision issues, clustering of images and object matching. The outcome of the state-of-the-arts in both applications is to be extensive.

2.4 GRAPH PARTITIONING

The partitioning heuristics are divided into global and local search methods. Global methods are sometimes called as construction heuristics for the reason that they take graph description as input and generate a balanced partition. Local search methods are called improvement heuristics. They take a graph and balanced partition as input and try to improve the partition. Each partitioning algorithm has to include a global search method. The Optimal method produces the optimal results, although they have exponential runtime behavior. The heuristic methods like Linear, Scattered and Random are exclusively dependent on the node position in the given node list.

A global search technique includes geometric partitioning algorithms, combinational partitioning algorithms, spectral partitioning algorithms, and multilevel-schemes. Local search techniques are node swapping local search, k-way local search, Tabu search, Genetic search, flow-based search.

Geomantic graph partitioning method can be processed based on the graph nodes coordinate information, not on the edges/links of the nodes. This is especially useful in finite element models and other geometrically-defined graphs from traditional scientific computing. Here, geometrically compact regions often correspond to graph blocks with a small cut. Partitioning using nodal coordinates comes in many flavors, such as coordinate nested dissection (CND) or recursive coordinate bisection (RCB) (Simon, 1991) and inertial partitioning (Williams, 1991; Farhat and Lesoinne, 1993). In each step of its recursion, RCB projects graph nodes onto the coordinate axis with the best expansion of the domain and bisects them through the median of their projections. The bisecting plane is orthogonal to the coordinate axis, which can create partitions with large separators in a case of meshes with skewed dimensions. Inertial partitioning can be interpreted as an improvement over RCB/CND in terms of worst-case performance because its bisecting plane is orthogonal to a plane that minimizes the moments of inertia of nodes.

In (Miller et al., 1991; Gilbert et al., 1998) of random spheres algorithm generalizes the RCB algorithm by stereographically projecting the d-dimensional nodes to a random d+1 dimensional sphere which is bisected by a plane through its center point. This method gives performance guarantees for planar graphs, k-nearest neighbor graphs, and other "well-behaved" graphs. Other representatives of geometry-based partitioning algorithms are space-filling curves (Pilkington and Baden, 1994, Hungershofer and Wierum, 2002, Zumbusch, 2003, Bader, 2013) which reduce d-dimensional partitioning to the one-dimensional case.

Space-filling curves define a bijective mapping from V to {1, …, |V|}. This mapping aims at the preservation of the nodes' locality in space. The partitioning itself is simpler and cheaper than RCB once the bijective mapping is constructed. A generalization of space-filling curves to general graphs can be done by so-called graph filling curves (Schamberger and Wierum, 2004). A recent work attempts to bring information on the graph structure into the geometry by embedding arbitrary graphs into the coordinate space using a multilevel graph drawing algorithm (Kirmani and Raghavan, 2013).

A combinatorial partitioning algorithm, adjacency information helps to portion a graph but on the non- coordinates of the nodes. For this explanation, the partitioning produced usually has minimum edge-cuts and is presumption to control disconnected sub-fields differentiating to those formed by geometric schemes. But, a combinatorial method tends to be time-consuming than geometric partitioning methods.

The Kernighan-Lin heuristic is the most frequently used local bisection method (Kernighan and Lin, 1970). It consists of a tiny number of passes throughout the nodes. In each pass, the algorithm constantly finds a pair of nodes, one from each of the sub-fields, and exchanges their sub-fields. The pairs are chosen so as to give the maximum upgrading in the quality of the bisection. Once a pair of nodes has been moved, the rest of the passes are considered for a move. The pass ended occurs all the nodes have been move. At this point, the minimum edge-cut was achieved by restoring the state of bisection. The input of the resulting bisection was performed by algorithm of the pass. The Kernighan-Lin (KL) algorithm usually acquires a small number of passes to meet. Every pass of the KL algorithm takes O (|V|2).

Fiduccia and Mattheyses (1982) introduced a modified KL-method that improves its run and use a sequence of a single node move to determine the sets. They also use a very efficient data structure called Buckets, which reduces the number of steps for each pass to only O (|V|+|E|). The basic essences, if Fiduccia and Mattheyses (FM) algorithm is same as KL algorithm, define gains for each node of the graph, select one node according to some criterion, remove it from its present partition and put it to the other partition, lock that node, update gains of all other unlocking vertices and iterate these steps until it reaches a local optimum configuration.

Diekmann et al., (1994) suggested a local improvement method which is a Helpful-sets heuristic is based on local rearrangements. The idea of Helpful Sets is to improve the cut size in several rounds, each of which consists of two steps. In the first step, a set is searched for one part and is moved over to the other part. Because the resulting bisection is not balanced, the second step searches for an equally weighted set on the over-weighted part and moves it to the under-weighted pair. The sets are chosen very carefully to force an improvement of the cut size after their exchange.

Kirkpatrick et al., (1983) suggested a mechanism based on local search based on statistical methods named as Simulated Annealing (SA). Johinson et al., (1989) have adapted simulated annealing to graph bisection and evaluate it with test results by using KL algorithm. Even though simulated annealing achieves better results than KL for few types of graphs.

Glover (1989 and 1990) suggested another general combinatorial optimization technique called Tabu search. Rolland et al., (1996) tabu search method was successfully used for partitioning a graph. The method recursively searches for good partitions by selecting random balance. During every call, a node is selected and moved from one component to another part until a balanced partition results with minimum cut size than the previously balanced partition encountered. If a founded partition was better after a fixed number of repeated moves, the algorithm raises an imbalanced factor. These factor restrictions the cardinality difference between the two divisions in the partition. The algorithm was experimentally compared with the SA and KL algorithms and found that is superior algorithm both with respect to quality and dynamically.

Spectral partitioning is one of the foremost methods to divides a graph into two subgraphs Spectral techniques were first used by Donath and Hoffman, (1972, 1973) and Fielder, (1975) and have been improved accordingly by others (Boppana, 1987; Pothen et al., 1990; Simon, 1991; Hendrickson and Leland, 1995; Barnard and Simon, 1993). Spectral bisection assumes for computing the eigenvector by global information of the connectivity of a graph with the second eigenvector of the Laplacian matrix of a graph. This eigenvector is also called as Fielder vector, it is the solution of a relaxed integer program for cut optimization.

The second eigenvector can be computed utilizing a customized Lanczos algorithm (Lanczos, 1950). Conversely, running time of this method is expensive. Barnard and Simon (1993) utilize a multilevel scheme to obtain a fast estimate of the Fiedler vector. The algorithmic structure is similar to the multilevel scheme, but their method coarsens with independent node sets and performs local improvement with Rayleigh quotient iteration. Hendrickson and Leland (1996) expand the spectral scheme to divide a graph into more than two parts by using multiple eigenvectors, these eigenvectors are computationally inexpensive to obtain. The scheme produces best partitions than recursive bisection, but this is especially useful for partitioning a graph with four or eight partitions.

Partitioning algorithms are implemented based on Multilevel Paradigm/archetype (Gupta, 1996; Hauck and Borriello, 1995; Hendrickson and Leland, 1995; Karypis and Kumar, 1998a, 1998b; Monien et al., 1999; Walshaw and Cross, ;). This archetype consists of three phases. The first phase is graph coarsening, which is a sequence of graphs is constructed by collapsing together selected nodes of the contribution graph for another cycle of graph coarsening until a tiny graph is obtained. The second phase is initial partitioning: computation of the initial bisection is executed on each of these graphs and so is very fast, and the third phase is multilevel refinement, it is executed on each level of the graph, from the coarsest to the finest. Multilevel archetype works for two causes. A good coarsening scheme can hide a large number of edges on the coarsest graph is first and an incremental refinement ideas such as KL and FM algorithm are more powerful in the multilevel framework.

Bui and Jones (1993) autonomously developed the multilevel paradigm is a fill-reducing matrix re-ordering computation. Hendrickson and Leland (1995) adopted multilevel paradigm and utilized in the environment of finite-element mesh partitioning. Hauck and Borriello (1970) used KL/FM scheme for optimization and hypergraph partitioning uses KL/FL by Cong and Smith (1993). Karypis and Kumar (1998a) studied this paradigm, by computing a selection of coarsening, initial partitioning and refinement in the graph context from different functions. Their experiment results present overall paradigm is relatively robust and constantly outperformed the spectral partitioning method in terms of both acceleration and quality of partitioning. Initial partitioning method is demonstrated for the coarsest graph that does not produce a quality solution. Initial partitioning of the coarsest graph is poor, the utilization of a local refinement scheme in the multilevel context executes in high-quality partitioning by and large. The evaluation also describes the use of the heavy edge matching and random edge matching in the partitioning of a graph coarsens, good results are obtained by using heavy-edge heuristic in multilevel refinement. This permits the utilization of really simplified variants of KL/FM schemes during the un-coarsening phase. This simplified scheme significantly accelerates refinement with quality partitions.

Karypis and Kumar (1998b) present a design for refining a k-way partitioning that is an overview of simplified variants of the KL/FM bisection refinement algorithm. k-way refinement scheme is utilized by Karypis and Kumar (1995), whose run time is linear in the number of edges (i.e., O(|E|)), whereas the runtime of the multilevel recursive bisection scheme is O (|E| log k). Experiments on a large number of graphs arising in various domains including finite-element methods, linear programming, VLSI, and transportation explain that this scheme produces partitioning are better quality than those produced by multilevel recursive bisection, and requires substantially less time.

A geometric methods, ie., Recursive Coordinate Bisection (RCB) and the inertial method are very fast in create partitions with high cut sizes evaluated with RSB. The geometric method developed by Gilbert et al. (1995) constructs good partitions when 30 or more random hyper planes are attempted at every bisection step.

Combinatorial partitioning algorithms such as RSB and KL recursive offers good partition with relatively slow. The multilevel approach as used in the multilevel-KL and multilevel RSB method outcomes in much speed without any decline in quality of the partition. Karypis and Kumar (1995c) statement that multilevel k -way algorithm provides better separations than multilevel –RSB. They compute a 256-way partition of a 44800-node finite element mesh in 40 seconds. A probable disadvantage is memory intensive in multilevel approach.

Recursive global methods are combined with some other local improvement method which leads to significantly minimizes cut sizes. Inertial and RSB methods offers cut sizes better but together with the KL method. Multilevel-KL method still provides as good cut sizes with less time for RSB-KL. Multilevel-KL and RSB-KL offers better cut sizes while compared with inertial-KL but slower than inertial-KL (Leland and Hendrickson, 1994). Methods depend on helpful sets do not look competitive with the multilevel-KL method (Diekmann et al., 1994).

2.5 FREQUENT SUBGRAPH MINING

In this section describing about Frequent Subgraph Mining techniques as follows

2.5.1 Inductive Logic Programming

Logic based mining, popularly known as inductive logic programming (ILP). ILP is characterized by the employ of logic for the representation of structural data. ILP systems represent instances, environment knowledge, hypothesis and target concepts in Horn clause logic. The core of ILP is the use of logic for representation and the search for syntactically legal hypothesis constructed from predicates provided by the background knowledge. ILP Systems such as First Order Inductive Logic (FOIL) (Quinlan, 1990), Golem (Muggleton and Feng, 1990), CProgol (Muggleton, 1996), WARMR (Dehaspe and Toiv6nen, 1999) and Faster Association Rules for Multiple Relations (FARMAR) (Nijssen and Kok, 2001) have been extensively applied to supervised learning and to a certain extent to unsupervised learning.

Dehaspe and Toivonen (1999) suggested a WARMR, which is a powerful Inductive Logic Programming (ILP) based algorithm discovers frequent queries. The Datalog origination is very common, as it permits the employ of the frequent pattern discovery problem. It consists of two-fold. First, exploratory data mining is well supported.

Next, the unified depiction gives the blurred picture of the frequent pattern detection. A number of dimensions appear Inside the Datalog origination that re-links move away settings. They demonstrated the frequent query methodology and utilized different applications, one analysis of alarm and other in a domain of chemical toxicology.

Nijssen and Kok (2001) proposed an efficient algorithm for discovering queries. Faster Association Rules for Multiple Relations (FARMER) which has been implemented as a modification of WARMR. A major problem of WARMR is that it heavily depends on a good implementation of θ-subsumption. This is prohibitive as θ-subsumption is an NP-complete problem (Kietz and Luibbe, 1994). The proposed algorithm achieves in efficiency by two properties of the WARMR algorithm are first using the first order logic notation; FRAMER removes the need for PROLOG. Second, by using a more sophisticated data structure borrowed from an implementation of Apriori, FARMER algorithm does not depend on a time-consuming test for equivalence. Experiments with FARMER algorithm then show a considerable speed-up compared to WARMR.

2.5.2 Apriori based Approach

Inokuchi et al., (2000) proposed an approach named as Apriori-based Graph Mining (AGM) to efficiently mine the associate rules among the frequently appearing substructures in a given large graph data set. It is a vertex based candidate generation method that increases the substructure size by one vertex at every call. Two frequent graphs of size-k are joined only when the two graphs have the same size-(k − 1) subgraph. Here the size of a graph means the number of vertices in a graph. The newly formed candidate includes the common size-(k − 1) subgraph and the additional two vertices from the two size-k patterns. The approach performance has been evaluated for the artificially simulated data and also for the carcinogenesis data of Oxford University and National Toxicological Program (NTP). The powerful performance of this approach under some practical conditions has been confirmed through these evaluations.

Kuramochi and Karypis (2001) proposed an efficient algorithm for discovering all frequent subgraphs (FSG) in a large graph dataset. The suggested procedure is edge-based scheme that raises the substructures size by adding one edge in every call of an apriori algorithm. Two size-k sequences are combined when they share the equivalent subgraph that having $k - 1$ edges is called the core. The newly produced candidate includes the core and the further two edges from the size-k sequences. Other than these two for each candidate has one more. Experiment analysis was done on synthetic dataset as well as a chemical component dataset which performed well for algorithm.

Huan et al., (2003) proposed an approach for Mining of Frequent Subgraphs in the existence of Isomorphism. Fast Frequent Subgraph Mining (FFSM) considers graphs as a matrix of triangle with vertices labels on the sloping, edge labels in another place. All the entries are concatenated to represent the matrix code, line by line from left to right. Based on lexicographic sequencing, similar graphs have the equivalent canonical code - Canonical Adjacency Matrix (CAM). Two new structures are generated; two matrices of fragments are joining in FFSM. FFSM also requirements a secured expansion function to the CAM pairing a new edge-node is added to the final node. FFSM generates the refinement and after refinement making, FFSM computes matrix lines to validates to identify a create matrix is in canonical form else it can be shrinking. FFSM saves all the essential changes to avoid open subgraph to test isomorphism. However, FFSM saves the similar nodes, and ignoring the edges. This helps in impetus up the join and expansion function since new segments are embedding, calculated by collection of nodes actions. As an important fault of the mechanism, FFSM need to identify the perennial of a supplementary collections are not canonical subgraphs.

Kuramochi and Karypis (2004) proposed a data structure known as GREW. GREW computes from a single input graph to discover frequently connected subgraphs. GREW is architected and implemented to work on a outsized undirected graph and to discover sequences subsequent to associated subgraphs that have embedding of a large number of vertex- disjoints. Their experimental analysis produces GREW is efficient and scalable by numerous tests using graphs directly originated from the real datasets. GREW's less system requirements allowed to implement efficient and effective policy that were implemted to identify a bulk set of patterns that can wrap either the network of recurring subgraphs. This recommends that GREW can be utilized as a building block for algorithms with applications to clustering in developing task-specific pattern discovery and classification.

Kuramochi and Karypis (2004, 2005) proposed two different algorithms which are Horizontal Single Graph Miner (HSIGRAM) and Vertical Single Graph Miner (VSIGRAM) to identify every frequent subgraphs in a huge sparse graph. The Breadth First Search (BFS) and Depth First Search (DFS) are two approaches respectively and each pattern is supported and overlap graph based Maximum Independent Set (MIS) measure are described by Vanetik et al., (2006). Numerous disparity types of the MIS measures, including approximate and exact MIS measures, were designed. Evaluation verified that these methodologies performed well for large graphs minng, although VSIGRAM was better than HISGRAM. The reason for the scalability benefit of VSIGRAM algorithm is that it remains track of the embeddings of the recurring subgraphs beside the DFS path, resulting in minimal subgraph isomorphism validation. The results indicated that SUBDUE executes worse than both HSIGRAM and VSIGRAM, SUBDUE have a tendency to spotlight on tiny subgraphs with high recurring and as a result tends to importants of missing patterns.

Huan et al., (2004) described a tree based SPanning maximal graph MINing also known as SPIN. Author proposed a mechanism that mines only maximal recurring subgraph; here subgraphs are not a part of any other recurring subgraphs. This may gradually decreases the size of the outcome set in the optimal case. It is the new FSM approach to mine frequent maximal subgraphs from a huge databases. The multiple optimization methodologies are included to expedite the method of mining.

Thomas et al., (2006) proposed MARGIN on the examination that the set of potential increase recurring subgraphs are integrated in the set of frequent k subgraphs that have infrequent (k + 1) super graphs. The set of candidates is frequently identified by the main algorithm, the post-processing operation are utilized to find the maximal frequent subgraphs. Evaluation outcomes visualized MARGIN is faster than gSpan when pertain to few graph databases. The initial cut increases the efficiency of MARGIN.

Thomas et al., (2009) suggested an algorithm called Item-set based Subgraph Mining (ISG). It discovers maximal frequent subgraphs for a database of graphs having unique edge labels using item-set mining techniques. In an item-set mining approach, an item set can be generated by using lexicographical order on the items. however, it is difficult to guarantee that a subgraph is generated only once during the process of frequent subgraph mining because a subgraph can be grown in several different ways by adding nodes or edges in different orders. Therefore, avoiding multiple explorations of the same graph add to the overhead of graph mining algorithms. The frequency of an item-set can be determined based on a Transaction IDentifier (TID)-list. The unique edge label property enables ISG to use maximal item-set mining algorithm to find the maximal frequent subgraphs. The running time of ISG and experimental results are compared with gSpan, Spin and MARGIN. ISG performs significantly better than the rest of the algorithms.

2.5.3 Pattern Growth Approach

Holder et al., (1994) proposed a Substructure Discovery in the SUBDUE system. Minimum description size approach applied to identify substructures, which uses the compressed database and in the data a structural concept is represented. Restoring previously identified substructures in the data, many passes of SUBDUE outcomes a ordering description of the structures in the data. The optimal environment knowledge guides SUBDUE towards suitable substructures for a exact domain. The use of an inexact graph matching allows a controlled amount of deviations in the instance of a substructure concept. A large amount of structural information that can be added to non-structural data collection on physical phenomena provides a large test bed for comparing an integrated discovery system based on SUBDUE to other non-structural system.

Borgelt and Berthold (2002) proposed a MoFa algorithm which is referred as Molecular Fragment Miner. The MoFa algorithm keeps all the list of embedding formerly identified subgraphs with an edge and the expansion function is constrained only to these embedding lists. Isomorphism tests can be done inexpensively by testing whether an embedding list can be polished in a similar way. The algorithm also uses structural shrinking and environment knowledge to decrease support computation and to remove duplicates using benchmark isomorphism testing.

Yan and Han (2002) discovered gSpan algorithm referred as graph-based structure pattern mining. gSpan a novel algorithm and new approach to Frequent based Pattern Mining. Without candidate generation gSpan is the initial algorithm that explores Depth First Search (DFS) in the frequent subgraph mining. This algorithm consists of two phases (i) DFS Lexicographic order (ii) minimum DFS code which forms a canonical labeling system to sustain DFS search. gSpan determines all the frequent subgraphs with required support. Since the design combines the subgraph isomorphism test and frequent subgraph growth into one procedure, gSpan dramatically hasten the mining process.

Yan and Han (2003) proposed CloseGraph algorithm referred as mining closed frequent graph pattern. It is the extension of the gSpan algorithm. A graph is closed in a database there is no proper subgraph is exist in graph that has the same support as X. CloseGraph is implemented by discovering several motivating looping methods. The CloseGraph performance not only exponentially decreases preventable subgraphs but also significantly the efficiency of mining is increase, particularly in large graph patterns.

Nijssen and Kok (2004) presented a GASTON algorithm which is referred as GrAph/Sequence/Tree extractiON. GASTON algorithm puts together frequent path, sub-tree, and subgraph, owing to the surveillance that most frequent substructures in molecular datasets are open trees. The algorithm suggests an explanation by segregate the frequent subgraph mining procedure into, the path, sub-tree, and then the generated subgraph. As a result, the generated subgraphs are appeal when it necessary. Hence, GASTON functions optimally when graphs are generally trees or paths because the most exclusive subgraph isomorphism testing is finding out subgraph mining phase. GASTON maintains all the embeddings in order to produce only new subgraphs, thus saving on unnecessary isomorphism detection. GASTON can compute the frequency of a subgraph either with isomorphism tests or embedding lists.

Han et al., (2007) have proposed an improvement in the gSpan subgraph finding algorithm, which is named Frequent Substructure Pattern Mining (FSP). The improvement recommended in gSpan algorithm has improved the subgraph identify: (1) Enhancing the canonical graph illustration used in gSpan and defining association between sub-DFS trees whose parent allocate the same parent node in the candidate child subgraphs; (2) Finding the structural relations in the DFS Search Space and representing the relation between child and siblings subgraph; (3) Demonstrating the two mechanisms into subgraph mining problem to minimize the number of graphs and subgraph isomorphism evaluation effectively; and (4) Designing the approach for FSP and alternative related mechanisms and validating against the outcomes.

Liu et al., (2009) describes about mining of repeated jump patterns from databases. The repeated jump pattern mining can radically decreases the number of graph patterns output, and still detain interesting graph patterns. The Jump Pattern Mining (JPMiner) an efficient algorithm, which incorporating the function of jump patterns verify into the DFS code tree. The experimental analysis of JPMiner using both real and synthetic datasets visualizes the number of frequent jump patterns are smaller than the closed frequent graph patterns, and also efficient and scalable in mining frequent jump patterns.

Lin et al., (2009) suggested a frequent subgraph algorithm, mSpan for directed labeled graphs based on a minimum code. Frequent Pattern-growth is adopted by an mSpan algorithm to get a minimum edge code and an abstract node code sequence is used to identify a graph sequence individually through minimum extension. It can also solve the isomorphic graph pattern and the redundant extension problem. The conduct test shows mSpan algorithm can find all the repeated patterns in database. However many patterns are found with one edge between two nodes are meaningless.

Winnie et al (2010) describe a technique for mining an interesting substructure from molecular database. It can identify interesting frequent subgraphs not only for the classification of different molecular structures but also for the characteristic of it from the others. Test statistic is utilizing, algorithm monitors every occurrence of subgraph to verify if they are interesting. An information theoretic measure shows the degree of interestingness. The interesting of each matched subgraphs from a total interesting measure for the unseen molecular structure classified into a specific class.

Hsieh and Li (2010) suggested a procedure for temporal information and form a connective subgraph of pattern mining. The algorithm is named as Temporal Subgraph Patterns algorithm (TSP). The TSP scheme in a depth-first search to recursively produces a pattern. The TSP algorithm needs to scan the entire database and does not generate unnecessary candidates, and hence low efficiency time and usage of memory. The TSP algorithm preserves efficiently mine closed patterns in a temporal Heterogeneous Information Networks (HIN) database. The experiment outcome show that the TSP-algorithm performs well on both synthetic and real datasets.

Tekin et al., (2012) suggested a graph mining based mechanism to notice similar design structures in object-oriented software architecture. This mechanism consists of three parts. First part is the model graph for implementation is created based on class diagram. Second step, is a partitioning algorithm to separate the software system into multiple parts and at last applying a subgraph mining algorithm to identify the similar design structures in the produced model graph. From this useful information can be obtained about the design, which might help designers to enhance their knowledge about the system design. They experimented on several open-source projects and found the results successful in indentifying uniform design structures in every project.

Tekin and Buzluca (2013) studied to get better design quality of design structure. They proposed a subgraph mining based method to identify uniform design structures in Object-Oriented Design (OOD) systems. Useful information regarding the design structures is obtained by classifying and analyzing these structures, which may be helpful for developers to gain knowledge that can be utilizing to creating a new design structures efficiently. They experimented on numerous open-source and industrial level projects and found successfully in finding several uniform design structures within the job and between different jobs.

2.6 CONCLUSION

In this chapter the latest research work done on software model graph representations of related topics are discussed that includes several concepts related to directed and undirected graphs, design patterns. Various other topics related to graph partitioning, graph clustering and frequent subgraph mining algorithms were studied.

The recent developments in various topics related to the proposed system are discussed. Several concepts related to software clones, reusability, and abstract syntax tree. Various papers related to software clones, reusability of source code and refactoring.

3. OVERVIEW OF PROPOSED METHODOLOGY

3.1 INTRODUCTION

Many information technology projects maintain an outstanding number of software replications through the software configuration management tools, which are identical or similar or duplicate copies of implementation or architecture references. Software clone concept is for various reasons such as reusing code by 'copy- and- paste' and others which make the source code to refactor. If faults found in a design or implementation, then the entire design model or source code needs refactoring and it is tedious for maintaining the quality as the software project implementation grows over a period of time. Software cloning is utmost important for project maintenance across release cycles and varying customer needs. Most of the software projects are frequently changing based on the business needs and end-user requirements and also improper use of design artifact can lead to poor quality and can increase the software budgeting. Therefore most of the software projects can consist of similar components and issues at the design level.

Domain Design Patterns are intended to solve domain specific problems with respect to the functional domain. These functional domains can be healthcare, finance, government, etc. Identifying the hidden patterns in a domain-specific approach is crucial for analyzing the architecture and to evaluate its quality. Identifying the design structures in a specific domain provides added advantages to solve the common issues within the domain.

In a software lifecycle, maintenance phase is highly cost-oriented, since most of the human resources need to have a thorough understanding of the architecture.
Maintenance phase involves fixing issues across several components of the designed system within the stipulated time and thorough understanding of architecture provides better feasibility to analyze the problem. Redesigning any of the components of the

existing software to improve quality and reduce the maintenance cost. Analyzing the architecture to identify the components that need to redesign is a crucial task, as the software developers need to understand the entire system behavior. During the re-architecture of software maintenance project, identifying which components or design in the system is identical. To improve the software quality and to reduce the cost of maintenance it is better to optimize and redesign frequently repeated identical design structures for improving quality of the software.

3.2 PROPOSED SYSTEM

In this section, a mechanism is suggested for finding similar design patterns in object-oriented systems is discussed using graph mining techniques. The proposed system is shown in Figure 3.1.

The proposed system is divided into of three phases. In the first phase, OOP software entities (i.e., classes and interfaces) and associations are extracted directly from the C# implementation by tokenizing the abstract syntax tree (AST). The builder component of the proposed system constructs the inheritance, association from the implementation and outcomes that is a complete graph of the software system. The proposed methodology supports C# implementation, the proposed methodology can be further extended to other OOP languages by parsing the syntax tree and generate the graph models. In the next phase, an entire software system is divided into multiple subcomponents by applying graph partitioning algorithm. These subcomponents are evaluated to identify similar designs within a project in the third phase of the proposed methodology

The outcome of the second phase i.e., subcomponents that are partitions are processed further to find similar design patterns. The graph partitioning and the subgraph mining algorithms, which are the critical parts of the proposed methodology, are further explained in detail in the subsequent sections.

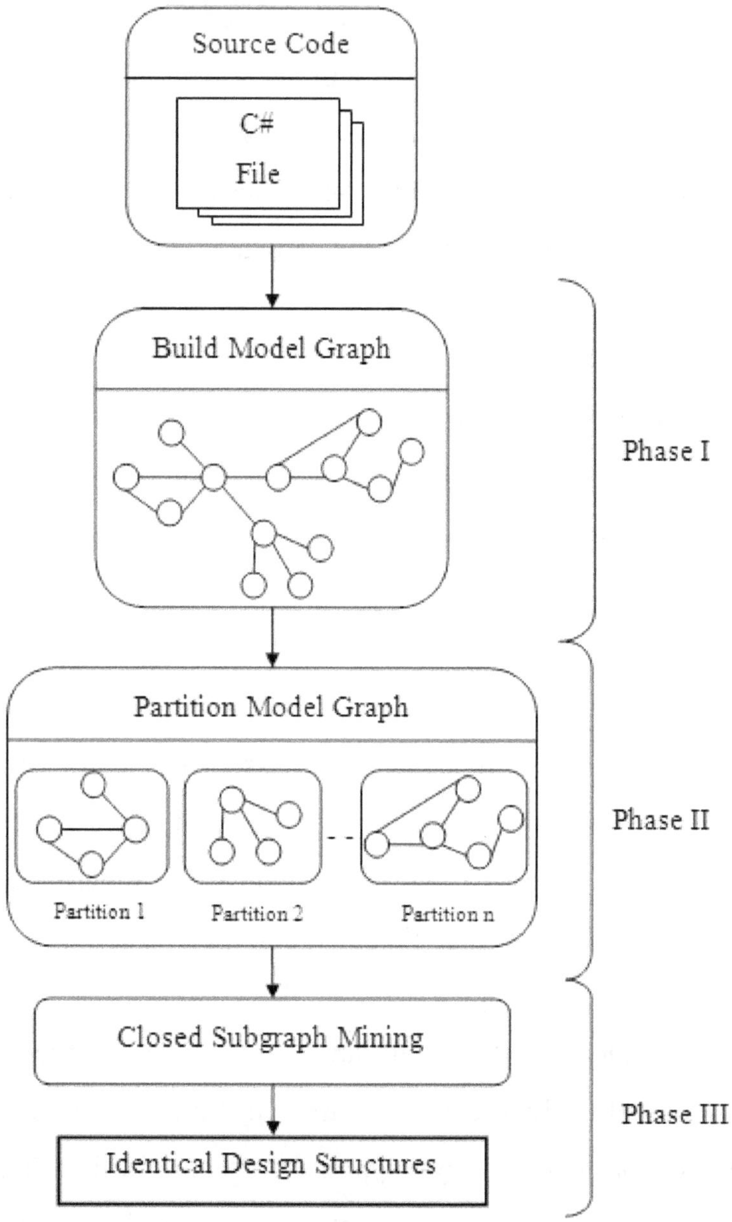

Figure 3.1: Block Diagram of the Proposed System

3.3 PHASE I: BUILDING CLASS-ORIENTED MODEL GRAPH

In this section, the construction of class-oriented model graph is described. The resulting model graph represents the software system (i.e. an Object-oriented model).

A model graph is built based on the classes and their associations between the various classes. In this construction, the NRefactor library is used to analyze the C# program code in a detailed view by parsing. The parsing contains syntax and semantic analyzer. The scanned program code will be visualized in a hierarchal tree in a unified console view. Few of the delegates are not been shown by syntax tree which is displayed by semantic analyzer by a type system. There is no mechanism for any of the programming languages implementation to analyze on its own, since most of the implementation is completely domain driven specific. Further to analyze these kind of systems, developers need to have an understanding of existing system components and objects interaction, etc, these type of information systems are known as type system. Every component is a collection of classes, these classes, in turn, consist of methods and attributes that describe the behavior of the real-time objects. The type system is of two types resolved and unresolved type system. The detail explanation can be seen in chapter IV.

3.4 PHASE II: PARTITIONING OF CLASS-ORIENTED MODEL GRAPH

In this section, a partitioning algorithm is proposed, namely Unfolding and Boosting based Graph Partitioning Algorithm (UB-GPA), which contains simple computations. It is an iterative process of two operations, unfolding and boosting until the stable matrix is formed resulting in a sparse matrix, which contains a maximum number of zeros in the matrix. From the stable matrix, the clusters are interpreted to graph partitions. The approach and the proposed algorithm is explained in detail chapter V.

3.5 PHASE III: IDENTIFYING FREQUENT IDENTICAL DESIGN STRUCTURES

This is the last phase of the proposed system; a closed frequent mining algorithm is implemented which is applicable to the multiple sets of graph data and the resultant graph obtained is a complete graph. The main aim of developing this algorithm is to find the identical design structures in a set of structures. The algorithm to find the closed related substructures is explained in detail in chapter VI.

4. PHASE I: GENERATING A CLASS-ORIENTED MODEL GRAPH FOR C# PROGRAMMING SOURCE CODE

4.1 INTRODUCTION

In Software Development Lifecycle (SDLC), maintenance phase is critical as this directly involves interaction with the customer and required solutions in a timely manner without any regressions. As the maintenance phase involves customer interaction and to provide fixes in a timely manner this phase is expensive, it is important to deliver high-quality software. Over the last decade, various investigations have been performed on source code to focus intricacy. This chapter proposes an approach for construction of Class Oriented Model Graph (COG) and analyzing source code using Abstract Syntax Tree (AST). The proposed source code analysis can recover the advanced structure of a software framework directly from its source code to produce a software model graph. Coupling, Common design structures, Refactoring design structures can be distinguished regardless of the fact that their implementation is altered. Refactored designs can be consolidated as an executable entity, to be utilized productively in various components of the software produced and deployed. The proposed methodology will likewise stay away from conflicting maintenance.

Implementing the design additionally assumes a key part in refactoring procedures, which includes receptive components to upkeep assignments. Redesigning enhances the inward architecture of the software by improving the quality and reducing the cost over a period of time. Recognizing these types of misused design patterns and recurring quality issues could offer a momentous provides benefits to the customers by reducing the cost; the reason is that irrelevant designs misused and can exploit the architecture for security and performance, scalability issues. For example, incorrect usage of design patterns can lead to erroneous and also result in a correct outcome, and common design quality issues can be immediately distinguished, allows to fix the issues across various components. In software, the design usually consists of reusable and frequent indistinguishable components of design structures, these components are right candidates in future design.

Most of the time, software engineer have a tendency to copy/past the implementation from various sources on the internet. Basically, seasoned software engineers modify these repeated components independently to permit their source code to change, yet the architecture remains intact (Li et al., 2006; Gandhi and Bhatia, 2011). The implemented software quality can deteriorate if developers provide a defect fix to one component and yet neglect to patch the fix in another amendment to its duplicates.

For a better quality of the software, usually the coupling of the design should be minimal i.e., the dependency of classes should be loosely coupled. The coupling provides an added advantage to overcome the difficultly and dependency issues when applying the defect fixes across the components. Modules within components should be highly cohesive to achieve the desired functionality i.e, the class is highly cohesive implies how attributes and methods are binded for the required behavior of the component. Coupling is constantly interrelated with cohesion, for a better quality of software highly cohesion and loosely coupled and vice versa (Saxena and Kumar, 2012). Coupling is taken into considering while designing the framework, it depends on the abstraction of the classes.

Let us now illustrate the cohesive class which can perform a solitary undertaking inside of the software component. It obliges little cooperation with various components that are utilized as a part of different components of software. Cohesion is used to decide how tightly bounded the internal attribute of a class to each other. Cohesion gives a proposal to the designer about whether the different attributes of a class have a place together in the same class. Thus, the Cohesion and coupling are associated with one other.

This chapter proposes an approach named Syntax Tree based Class Oriented Graph (COG-ST), to generate the syntax tree NRefactory open source implementation is used to retrieve the information at different components of object-oriented language (C#). The use of generating software model graph is to provide in-depth information regarding the interactions between the classes (vertices/nodes). The communication message of the nodes is shown with the edge weight in model graph, which helps developers in Refactoring, reused designs, common design structures, circular dependency structures and pattern design. All those factors can observe by utilizing COG- ST.

4.2 PROPOSED APPROACH FOR GENERATING MODEL GRAPH

The proposed approach consists of various steps, preliminarily considers a C# implemented file as input and then peruses all the tasks inside of that implemented file and then parses the task files (project files) to discover all the implementation inside the tasks. Amid iterate of different implementation present, it will likewise shift through the records that are checked for avoidance amid parsing. After all the required documents are perused in the framework, these are provided as input to language parser for constructing syntax tree. The parser considers C# implementation as an object. The object is further parsed and produces a syntax tree tells us that it is an IExpression, which has an association as an object.

```
IExpression
    {
        Tgt = new MemberRefernceExpression
        {
            Tgt = new IExpression("Object")
            attributeName = "DoIt"
        }
```

```
}
```

Complete information can't be retrieved from the syntax tree with respect to the specific object or DoIt. DoIt undoubtedly is an object method, and some object seems to be a parameter local to the present class. It might be that some object could be a name of the class and DoIt acts a delegator. The delegates are very useful in some cases like passing methods as parameters, callback, calling multiple methods in a single event, etc. the semantic tree offers the information with respect to these properties.

```
new IResolveResult
{
        TgtResult = new LResolveResult(" MyClass Object")
        Member = void MyClass.DoIt()
}
```

Once syntax tree and semantic tree are created, then the class-oriented model graph is built. The block diagram of the suggested approach Syntax Tree based Class-Oriented Model Graph (COG-ST) is shown in Figure 4.1. The reset of the things is explained as follows.

The open source NRefactory library is utilized to build Abstract Syntax Tree (AST). NRefactory analyzes C# source code implementation and constructs AST. NRefactory offers Application Program Interface (API) for retrieving syntax tree, semantic tree data, and the type system.

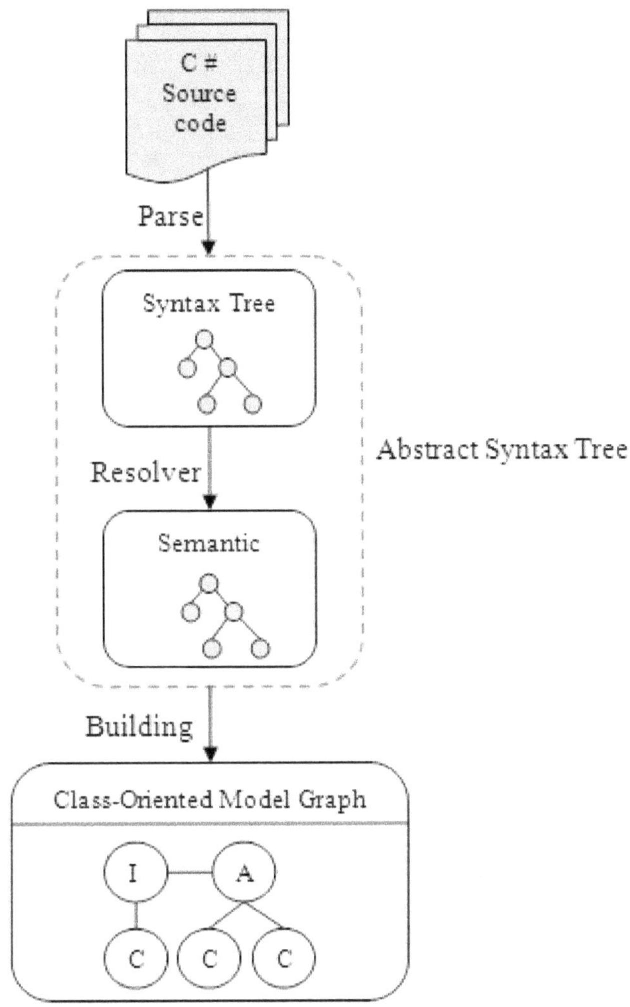

Figure 4.1: Block Diagram of Class-Oriented Model Graph

4.2.1 Syntax Tree

To parse C# source code, the below statement is utilized as CSharpParserclass:

```
CSharpParserclass parser = new CSharpParserclass ();
SyntaxTree syntaxTree =Par ser(programcode);
```

Parser.Errorscollection is used while parsing to retrieving syntax error detection. The syntax tree can be visualized in Figure 4.2 for the below sample source code

```
using System;
using System.Ling;
class sample
{
        public int Main(string [] args)
        {
                return args[0]. IndexOf(args [1]);
}
}
```

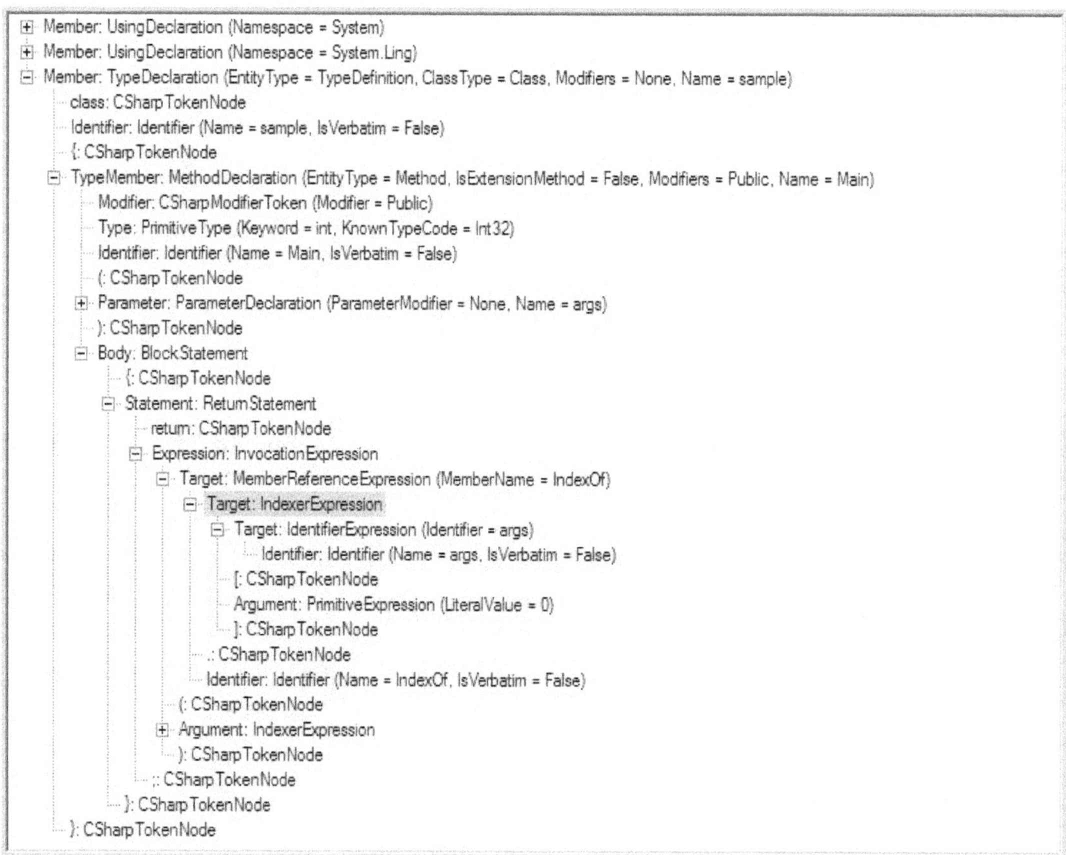

Figure 4.2: syntax tree

Abstract Syntax Tree Node (ASTNode) is the abstract class of the syntax tree and every member in the tree view is one ASTNode. Each treenode has a catalog of children and every sibling has a task. A sibling node task depicts the association between the parent node and the sibling node, it provides the explanation about where the parent node appears in.

The title describing the nodes in the tree view follow the pattern Role-node-getType(). In Figure 4.2, depicts the selected IndexExpression-args[0] has the sub-node identifier "args" in the objective task and the child "0" in the Argument role. For multi-dimensional arrays, there would be multiple siblings in the argument role which are divided by comma tokens.

However, the syntax tree is rather inconvenient filtering the children by role gets rather long-winded if you need to do it all over the place. Also, it is not always clear which tasks can be appearing in a given construct, below is the implementation details of the API

```
var args = siblings, Where(c => c.role = =Argument).Cst<Expr>();
var args  = indexer.Get siblings ByRole(roles.args);
var args = indexer.args;
```

4.2.2 Type System

Analyzing of C# source code file is not possible by its own. It needs to recognize which members and classes exist. This depends on other implementation in the same task, and on the associated assemblies and assignments. NRfactory library helps to gather all this data collection is known as the type system. Type system restrains not only types but also attributes events, methods, and behavior.

The type systems are of two different types in NRFactory, these are unsolved and resolved type. The unsolved type system is deliberated as a kind of ground staging for constructing a type system. The similar information contains in the unresolved type system and the syntax tree, however, it utilizes language-independent interfaces for representation. The unresolved type system does not hold any information in relation to method bodies and is much additional memory-efficient than keeping all around syntax trees

The below example consists a minimum project model that contains the classes solution, CpProject, and CpFile. For each project, MSFTBuild is used to open the extension file with .csproj file and finding out the compiler settings. CpProjectContent construct an object for each module, this is the parent instance of the unresolved type system.

For each implementation in the module, CpParser is invoked for parsing, then included it to the module content as follows:

```
cp = cp.AddOrUpdateFiles(sTree.ToTypeSystem());
```

The example described above also persists the complete syntax tree in the CpFile class. This makes the example application simpler, there can be an increase in usage of the memory when loading a larger application. In the Sharp Develop environment, we hold the syntax tree of the currently open file in memory. For the remaining files, keep minimum system type around

As for the referred libraries, the MSFT.Build library is used to invoke the MSFTBuild ResolveAssemblyRefernces job. This job determines the complete location to the library files. This will provides us the correct version of the libraries based on the module's target design.

To stores those libraries in memory, the CecilLoder class are utilized:

```
For (string assembly_File : ResolveAssemblyReferences(file))
{
IUnresolvedAssembly library = new
CecilLoader().LoadAssemblyFile(assembly_File);
mc=mc.AddAssemblyRefernces(library);
}
```

The last task, after the module has been loaded into memory, is to construct the type system of resolved:

```
compn = mc.constructCompn ();
```

In the description above, there is no need to consider module references. The class CpProjectContent implements IAssemblyRefernce, so it is feasible to construct a module reference by parsing module content to AddLibraryRefs similar to the library file references. There is no approach or mechanism to find out whether the referenced module already loaded- it might come later in the list of modules. And since CpSharpModuleContent is immutable, using module content directly would reference that exact version of the module content- this means it is impossible to create cyclic references this way.

4.2.3 Semantic Analysis

Armed with a type system, we have got all the knowledge we want to perform semantic analysis. To the class, CpAstResolver is using to discover the semantics from an AST node of C# implementation.

```
CpAstResolver resol = new CpAstResolver (compn, sTree, unresolvedFileType);
```

```
IResolveResult result = resolv.Resolve(ref);
```

The resolved type system and the parent node of the syntax tree implemented in C#, information are used to construct a C# AST resolver. By invoking the IUnresolvedFile references with the syntax tree and also registers in the type system was constructed. Resolve is offered to identify the association between the Syntax Tree declarations and the type system. Alternatively, to figure out this mapping the resolver will utilize the member signatures. In case there are defects in the C# implementation then it tends to be slower and may fail. So it is a decent plan to store the IUresolveredFile once making the file system, so that it can be provided to the resolver.

After creating a resolver instance, the Resolver () method is utilized to determine the semantics of any node within the syntax tree. The CSharpAstResolver has an internal cache of already resolved nodes, so if you use the same CSharpAstResolver to resolve both Indexof() calls in the following example, the type of the variable temp will be resolved only once.

```
Var temp= x.complex().expr;
integer x = temp.indexof("x");
integer y= temp.indexof("y");
```

An IResolveResult is a list of tree node instances that represent the expression of semantics. For following sample code the semantic tree can view in Figure 4.3

```
using System;
using System.Ling;
class sample
{
        public int Main(string [ ] args) {
                Console.WriteLine(args.Length * 1.5); }
}
```

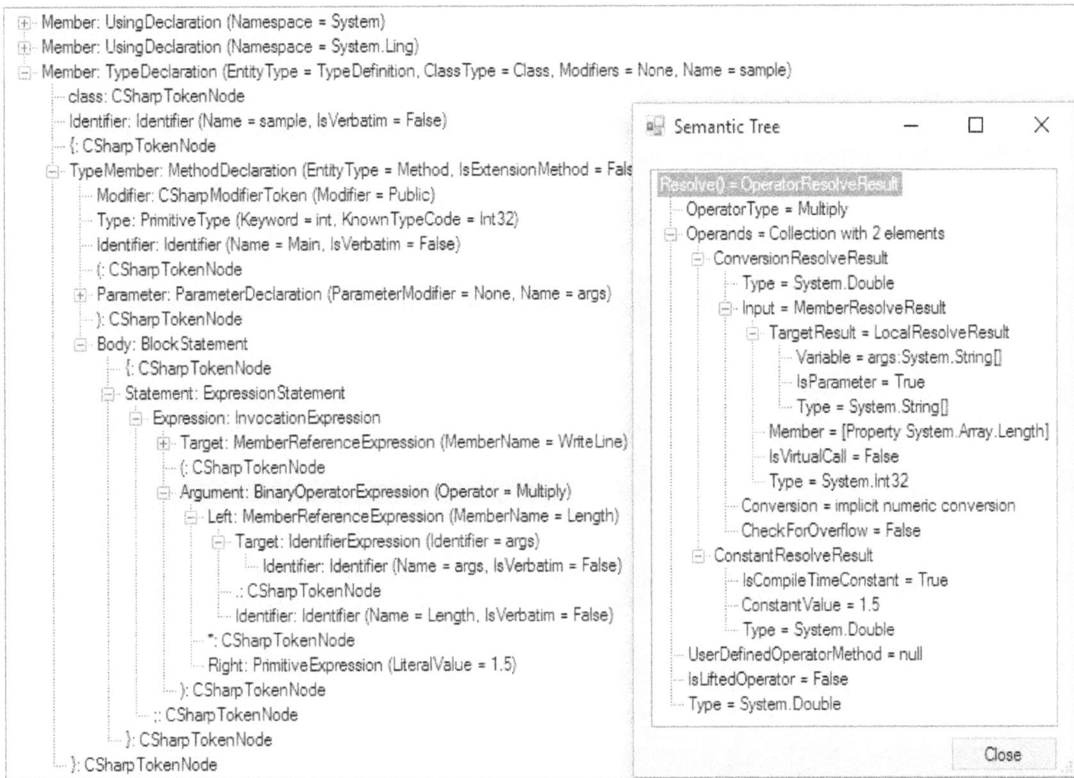

Figure 4.3: Semantic Tree for Example

The methods CpAst Resolver.Getconversion () and GetExpecteType() can be utilized to recall the change happened in the given node.

Finally, the CpAstResolver also provides methods that return the resolver's state at a given AST node. The next, CpAstResolver is utilized to evaluate a small piece of the code segment in the framework of a dissimilar file.

4.2.4 Class-Oriented Model as a Graph Representation

This section explains the construction of the class-oriented model graph, which represents the complex outlook of system architecture as an undirected label graph. The class, interface, and abstract classes are defined as nodes. Associations among the classes are defined as edges of the graph.

Various association types in a class-oriented graph depend on UML relations. Here, class diagram and sequence diagrams of UML are specially considered. Additionally, few unseen associations in the UML notations are handled here. For instance, a class has the same method name with the parent class method class, and then there is an override association between these classes that are not directly visualized in UML class diagrams. From UML sequence diagram, few main advanced relations are considered such as method call and parameter passing associations between classes, feasible entity types, association types and their edge labels are given in Table 4.1

Table 4.1: Vertex and Edge Labels of the Class-Oriented Model Graph

Vertex labels	**Class types**
A	Abstract class
C	Class
I	Interface
Edge labels	**Relation type**
A	Class U contains member variable of Class V
F	Class U access field of Class V
I	Class U implements Class V
L	Class U method has a Class V as a local parameter
M	Class U has invokes Class V through the method call
O	Class U overrides method of Class V
P	Class U in its method argument of Class V
R	Class V as the return types from Class U methods
X	Class U extends Class V

An Object-Oriented design, can have multiple associations between the entities (vertices). An understandable and simple graph is constructed with all the labels of edges are connected among the two classes (vertices) into a distinct set of labels with the pre-defined deterministic order. For illustrate, if two classes have both method parameter {P} and method call {M} connections in the same association, then the collective label set for this edges become {P}U{M}= {PM}, if the method parameter relation {P} come initially in the pre-defined sequence. The pre-defined sequence uses to guarantee that the combination of the similar associations gets always the label {PM} and not {MP}. In proposed mechanism to identify design level duplicates, taken into an examination of the entire label match.

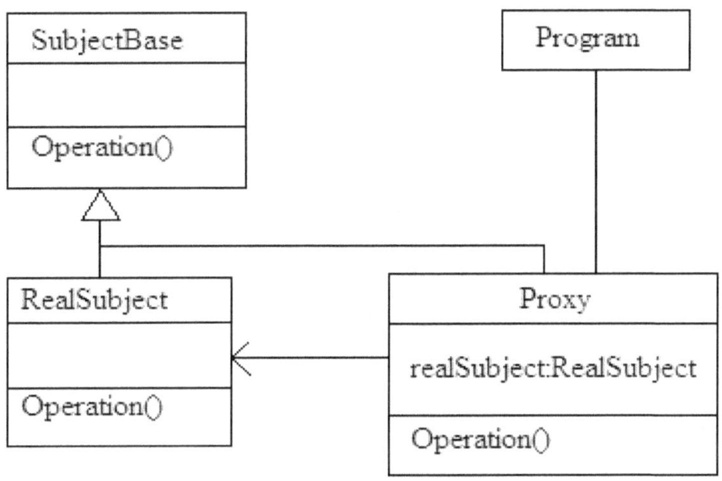

Figure 4.4: UML Class Diagram of a Sample Proxy Example

Figure 4.4 exhibit the class diagram of a proxy design pattern example and Figure 4.5 represents the class-oriented model graph was constructed. The class-oriented model graph is compared to a class diagram which holds additional information, such as local parameter {L}, method call{M}, extend {X} and field type {A}.

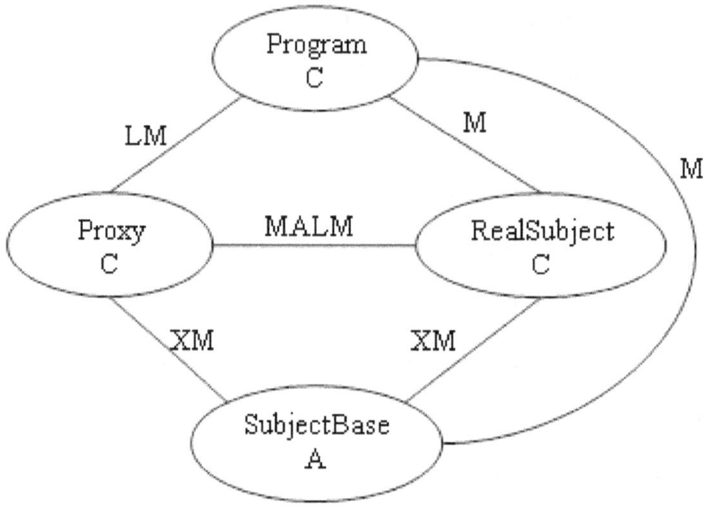

Figure 4.5: Class-Oriented Model Graph of the Proxy Pattern

4.3 PROPOSED SYNTAX TREE BASED CLASS-ORIENTED MODEL GRAPH (COG-ST) ALGORITHM

Input: C# Source Code

Output: Class-Oriented Model Graph

Step 1: Scan the entire source code and build the Abstract Syntax Tree (AST)

Step 2: AST is utilized to govern the semantics of any class nodes inside the AST.

Step 3: Identify the classes and their labels as

(i) Abstract class is labeled as A

(ii) Interface class is labeled as I

(iii) Normal class is labeled as C

Step 4: Identify the associations between classes and their labels as

(i) Extend association as an edge with X as edge label.

(ii) Implement association as edge with I as edge label

(iii) Field type association as edge with A as edge label

(iv) Local variable association as edge with L as edge label

(v) Parameter association as edge with P as edge label

(vi) Return type association as edge with R as edge label

(vii) Method association as edge with M as edge label

(viii) Override association as edge with O as edge label

Step 5: Represent classes as node/vertices and associations as edges with labels.

Degree of Coupling

The degree of coupling (Dakkal, 2015) is calculated as the proportion of a number of messages received to the number of a message produced. The degree of coupling is given in equation 1. DC represented the degree of coupling, MRC is the message received coupling and MSC is message sender coupling. The MRC computes the difficulty of the message accepted by the classes, as MRC is the number of messages accepted by a class from the other classes. The MSC is the number of message sender coupling among the instances of the classes. The low- level coupling is attained during the message between the components.

$$\text{Degree of Coupling (DC)} = \frac{\text{Number of Message Received Coupling}}{\text{Number of Message Sender Coupling}} \qquad 4.1$$

4.4 ILLUSTRATION OF CLASS-ORIENTED MODEL GRAPH

According to the algorithm of the class-oriented model, the graph is explained by observer pattern. Here, the input is the sample program called observer patterns. The observer design patterns define a connection between instances so that when one instance state changes, all needy instances are updated automatically. This pattern allows communication between objects in a loosely coupled manner.

The sample source code given below is loaded into the parser. Then it generates the syntax tree. If any delegate functions which are not defined in the syntax tree, by using resolve function it is going to elaborate the tree in semantic.

Pseudo code for observer pattern

```
public class Subject {
        public Observer o;
        public void SubjectMethod() {}
}
public abstract class Observer {
        public Subject s;
        public void ObserverMethod( Subject s );
}
public class Observer1 : Observer {
        public void ObserverMethod( Subject s ) {
                s.SubjectMethod();
        }
}
public class Observer2 : Observer {
        public void ObserverMethod( Subject s ) {
                s.SubjectMethod();
        }
}
public class Observer3 : Observer {
        public void ObserverMethod( Subject s ) {
                s.SubjectMethod();
        }
}
```

The generated syntax tree is shown below:

The above sample program is loaded into parser then it generated the syntax tree which is given as follows

+ Member: TypeDeclaration (Entity Type = TypeDefinition, Class Type = Class, Modifier = Public, Name = Subject)

+ Member: TypeDeclaration (Entity Type-TypeDefinition, Class Type = Class, Modifier = Public,Abstract, Name = Observer)

+ Member: TypeDeclaration (Entity Type-TypeDefinition, Class Type = Class, Modifier = Public, Name = Observer1)

+ Member: TypeDeclaration (Entity Type-TypeDefinition, Class Type = Class, Modifier = Public, Name = Observer2)

+ Member: TypeDeclaration (Entity Type-TypeDefinition, Class Type = Class, Modifier = Public, Name = Observer3)

From the above syntax tree, one abstract and four classes are observed. All those classes are bold in the sample source code. The hieratical form of the class subject is shown in Figure 4.6.

Figure 4.6: Syntax Tree of Sample Program for Observer

public class **Observer1: Observer**

The line defines that observer1 is derived from the Observer abstract class. As per table 3.1, it defines the association between these two classes as X.

```
public class Subject
{
        public Observer o;
        public void SubjectMethod() {}
}
public abstract class Observer {
        public Subject s;
        public void ObserverMethod( Subject s );
}
```

From the above pseudo code, the associations between subject and observer and their edge labeled are A and AP. The link between subject and observer is field type (A) because in subject class there is an observer object 'o'. Similarly, in observer class there is a subject object 's' is defined and the Subject s is passed as an argument in the ObserverMethod( ) so it has parameter passing (P).

```
public class Observer1 : Observer
{
        public void ObserverMethod( Subject s )
{
                s.SubjectMethod();
        }
}
```

From above pseudo code Observer1 is derived from abstract Observer so it's association label is X and in Observer1 having SubjectMethod( ) and Subject is passing as a parameter so it having the association of M and P (MP). Similarly, Observer 2 and Observer 3 having the same associations and their edge label.

By using Boosting Graph Library, the model graph is visualized in Figure 4.7

In the array, it stores all the class labels as

1	2	3	4	5
Observer 3	Observer2	Observer1	Observer	Subject

1→4 5 (Observer3 is linked to Observer and Subject)

2→4 5 (Observer2 is linked to Observer and Subject)

3→ 4 5 (Observer1 is linked to Observer and Subject)

4→5 1 2 3(Observer is linked to Subject, Observer3, Observer 2 and Observer 1)

5→4 1 2 3 (Subject is linked to Observer, Observer3, Observer 2 and Observer 1)

This shows the connectivity of the entities in the model graph and their edge label is stored and displayed on links.

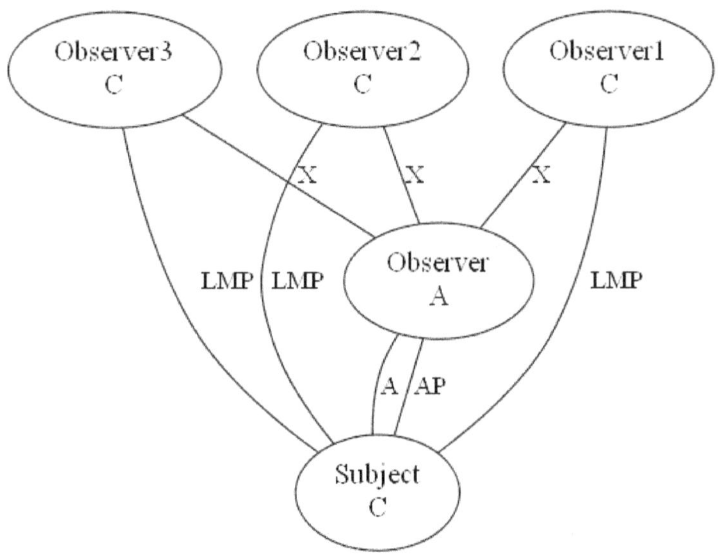

Figure 4.7: Undirected Label COG- ST

The class labels and edge labels of the graph are follows are stores in the note-pad for the further processing.

```
graph G {
node [fontsize = 16];
Observer3 [label="Observer3\nC"];
Observer [label="Observer\nA"];
Subject [label="Subject\nC"];
Observer2 [label="Observer2\nC"];
Observer [label="Observer\nA"];
Subject [label="Subject\nC"];
Observer1 [label="Observer1\nC"];
Observer [label="Observer\nA"];
Subject [label="Subject\nC"];
Observer [label="Observer\nA"];
Subject [label="Subject\nC"];
Subject [label="Subject\nC"];
Observer [label="Observer\nA"];
Observer3 -> Observer [label=X];
Observer3 -> Subject [label=LMP];
Observer2 -> Observer [label=X];
Observer2 -> Subject [label=LMP];
Observer1 -> Observer [label=X];
Observer1 -> Subject [label=LMP];
Observer -> Subject [label=AP];
Subject -> Observer [label=A];
}
```

The generated model graph is directed label graph is depicted in Figure 4.8, then compute degree of coupling (i.e., MSC and MRC). The figure 4.9 gives information about observer 1class, where observer 1 class is communicating to subject class with three different messages, one is Local parameter (L), second is method (M), other is parameter (P) and to observer class with Extend (X). It performs coupling is MSC similarly rest of the class shown in figure 4.9 with labels (b), (c), (d) and (e). In figure 4.9 with label (d) communicates of MSC there is multiple labels (AP) on edge with reference two communications are performing. Table 4.2 describes about total number of MSC and MRC of various classes and their degree of coupling.

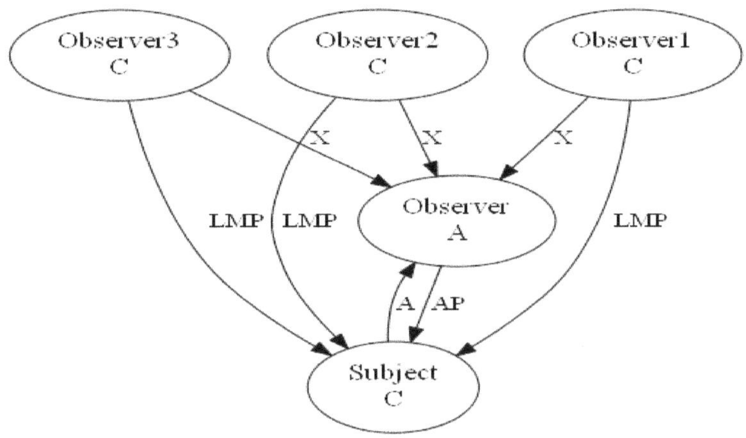

Figure 4.8: Directed Label COG-ST

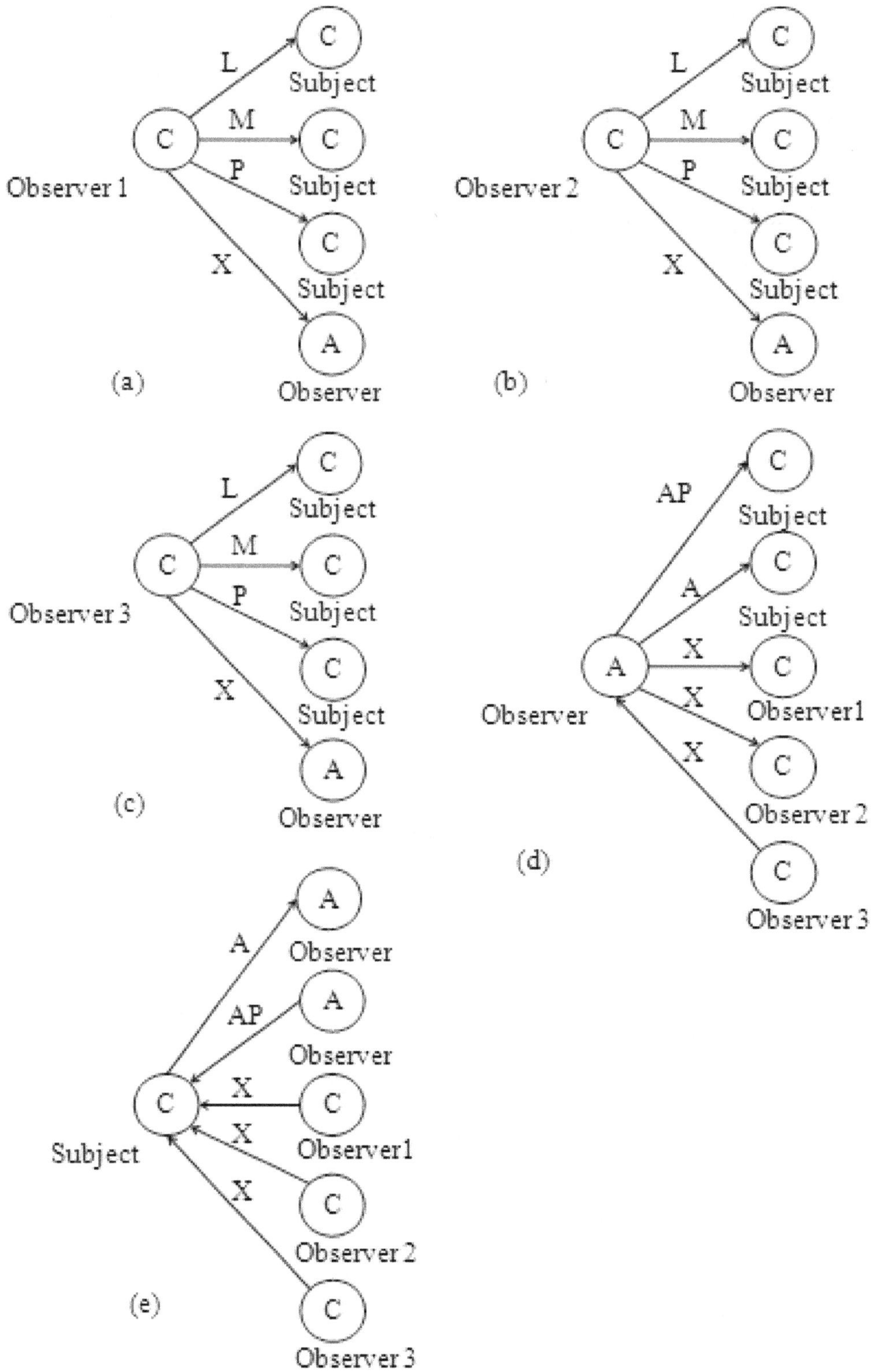

Figure 4.9: Degree of Coupling

Table 4.2: Coupling Matrices

Class	Object Oriented		
	MSC	MRC	DC
Observer 1	4	0	4/0
Observer 2	4	0	4/0
Observer 3	4	0	4/0
Observer	4	4	4/4 =1
Subject	4	4	4/4=1

4.5 CONCLUSION

In this chapter, an approach is proposed to generate a class-oriented model graph and to analyze the implementation using AST method. In suggested approach, the executable file for the C# application is taken as an input to the system. The file loads into memory and reads one after another to batch all the implementation in an array to create the syntax tree. Abstract syntax tree creation is completed, then ready for analysis. The NRefactory library is utilized to generate syntax tree. Generated syntax tree can be used for finding refactoring of similar source code and for finding a patterns design. The construction of class-oriented model graph provides in-depth information about a system. These model structured graphs can facilitate developers in better awareness on the architecture of the system.

5. PHASE II: PARTITIONING OF CLASS ORIENTED MODEL GRAPH

5.1 INTRODUCTION

Software systems are hard to comprehend the functionality of the large systems because systems are implemented with concepts of object-oriented programming. Therefore, it becomes difficult to extend and modify the software systems for the new application. Thus, the large and complex software systems can be divided into smaller components that are easier to understand by the software developer or designer.

The concept of breaking a large software system into small components can be achieved through clustering. The study of clustering is an essential and useful unsupervised learning method (Agrawal et al., 1998; Ester et al., 1996; Gibson et al., 1998; Ng and Han, 1994).The target of clustering is to cluster similar objects into one cluster while partitioning dissimilar instances into different clusters. Clustering has wide applications including the analysis of biological data, economic data, time series data and spatial data soon.

A graph is a meaningful data structure, widely used to represent a structural association between objects in many domain applications such as social networks, biological networks, web and fraud detection, so on. Graph clustering is challenging and an interesting research work carrying in recently (Bader et al., 2012) a large graph clustering tries to partition the graph into numerous compactly associated components. Usual applications of graph clustering include community exposure in social networks, classification of functionally connected protein structures in big protein-protein communication networks, etc (Schaeffer, 2007).

Software clustering is an important feature to divide a software system into smaller subsystems. It helps to discover the software subsystems that are interrelated functionally and independent from each other part of the system. Therefore, the re-engineering processes need to identify the subsystem of an activity to concern with the regeneration and improvement of the existing software application.

Designing reusable software (i.e. modular) system is an important principle in the Object-oriented domain. A modular programming system is a single unit of the application (software) architecture. Strongly coupled modules are encapsulated with a tiny number of classes. Modules that are interrelated with each other through a simple interface.

5.2 PROPOSED APPROACH FOR PARTITIONING A GRAPH

Author has suggested a new graph partitioning algorithm by adopting few features of MCL algorithm. In this algorithm symmetric matrix M is formed by combining adjacency matrix and degree matrix of G. The Min-Max normalization technique is applied to matrix M for transforming the data to a specified small range. The algorithm consists of two main operations namely unfolding and boosting which are implemented on the normalized matrix. These operations are repeated alternately until the stable steady matrix is reached. Finally, the clusters are interpreted from stable matrix S (i, j), where $1 \leq i \leq n$, $1 \leq j \leq n$, and n is a number of vertices. For the rows having the value one, the corresponding j indices are taken into one group called a cluster. Thus in this way, the clusters are interpreted from the stable matrix. The block diagram of the proposed approach UB-GPA is shown in figure 5.1.

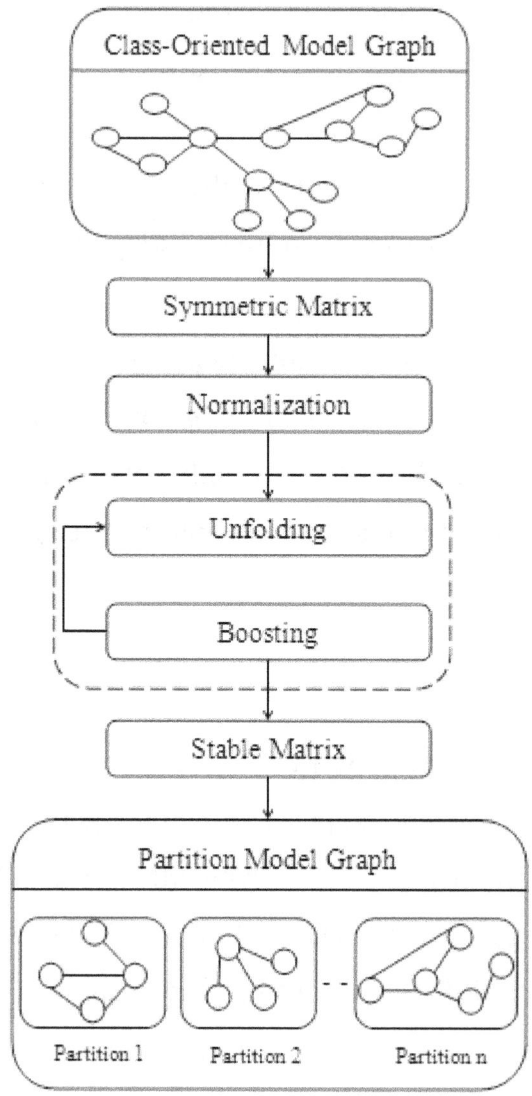

Figure 5.1: Block Diagram for the UB-GPA Algorithm

5.2.1 Symmetric Matrix of a Graph

The input for the second phase is an undirected weighted graph G, unfold parameter 'm' and boost parameter 'b'. From the input graph G, an adjacency matrix A is created and is represented as A= $[a_{ij}]$, where a_{ij} is the weight of the edge e_{ij}. Degree matrix D of graph G is found, which is represented as D= $[d_{ij}]$ and is defined as

$$d_{ij} = \begin{cases} \sum_{k=1}^{n} a_{ik} & \text{if } i = j \\ 0 & \text{otherwise} \end{cases} \quad 5.1$$

Adjacency and Degree matrix of G are combined to obtain the symmetric matrix M.

5.2.2 Normalization

The resultant symmetric matrix M is normalized by min-max normalization technique for transforming the data to a specified small range. The Min-max normalization map a value v to v′ in the range $[new\_min_X, new\_max_Y]$ for the attribute A, by computing

$$v' = \frac{P}{Q} * [new\_max_X - new\_min_Y] + new\_min_Y \quad 5.2$$

Where P= [v-minimum value of X] and Q= [maximum value of Y-minimum value of X]

In the proposed approach the minimum and the maximum value of the column of a matrix M is taken and the new range of the matrix is defined from [0, 1]. Thus the above equation for the matrix M (i, j) can be written as

$$v' = \frac{X}{Y} * (new\_range\_max - new\_range\_min) + new\_range\_min \quad 5.3$$

Where X = [v-min value of jth column], Y= [max value of jth column – min value of jth column] and $1 \leq j \leq n$, n is number of vertices

5.2.3 Unfolding Operation

After normalizing the matrix two main operations, namely unfolding and boosting are applied. These operations are repeated iteratively until a convergence is reached i.e. until a stable matrix is reached.

The Unfolding matrix M_U is achieved via matrix squaring and is defined as

$$M_U = M'_{ij} = \sum_{k=1}^{m} M_{ik} M_{kj} \quad 5.4$$

Where 'm' is an integer taken as 2. Clusters of large size are formed with the higher value of m. This may lead to a single cluster with no partitions and lower value of m (i.e. m=1) may lead to the problem of multiplication. Therefore m value is considered as 2.

5.2.4 Boosting Operation

The boosting is achieved by element square of the unfold matrix M' and element squared matrix is normalized. The boosting operation on the present state is responsible for both strengthening due to element squaring and weakening due to normalizing. Therefore to control the extent of the strengthening/weakening, boosting parameter b is considered. The low boosting parameter is more prone to yield smooth partitions.

$$M_B = M''_{ij} = \frac{(M'_{ij})^b}{\text{normalization}} \quad 5.5$$

Where, b is an integer value taken as 2.

The unfolding and boosting operations are iteratively repeated until it results in a stable state matrix.

5.2.5 Interpret Clusters

Finally, the clusters are interpreted from stable matrix S (i, j), where $1 \leq i \leq n$ and $1 \leq j \leq n$, n is a number of vertices. For the rows having the value one, the corresponding j indices are taken into one group called a cluster. Thus in this way, the clusters are interpreted from the stable matrix.

The key benefits of the proposed UB-GPA algorithm are (i) It is fast and scalable (ii) Its formulation is simple and elegant. (iii) It needs a minimum of iteration to produce stable state (i.e. Less than 15 iterations). (iv)It produces good clustering results.

5.3 PROPOSED UNFOLDING AND BOOSTING BASED GRAPH PARTITIONING ALGORITHM (UB-GPA)

Input: Software model graph G, unfolding parameter 'm', boosting parameter 'b'
Output: Software partitions

Method:

Step 1: Create an adjacency matrix A of G

Step 2: Find the degree of the matrix D of G

Step 3: Combine the adjacency and degree matrix to form a symmetric matrix M i.e. M = D+A

Step 4: The symmetric Matrix M is normalized M_N using min-max normalization technique.

Step 5: Apply simple matrix multiplication $M_N \times M_N$ to obtain unfolding matrix M_U.

Step 6: Apply element squaring to obtaining unfolding matrix and min-max normalization for the element squared matrix to obtain boosting matrix M_B.

Step 7: Repeat step 4 and step 5 until a stable state matrix M_S is obtained.

Step 8: Finding the number of ones in a row and combine all the indices into one group to form a cluster from stable matrix generated in step 7

In another way we can modify the algorithm, by using min-max normalization technique data can be transformed into a specific range. In boosting operation consists of two things one is squaring the elements and other is normalization. In this normalization technique, we can use column normalization (i.e, the summation of column =1). By changing the algorithm like this, there is no change in the results, but the iterations will increase in this operation while comparing with the UB-GPA. The above-mentioned changes in (Muttipati et al., 2015) with another way of constructing a model graph and applied.

5.4 ILLUSTRATION OF THE PROPOSED UB-GPA ALGORITHM

Let us consider an undirected weighted graph shown in Figure 5.2, which is an input for the UB-GPA algorithm.

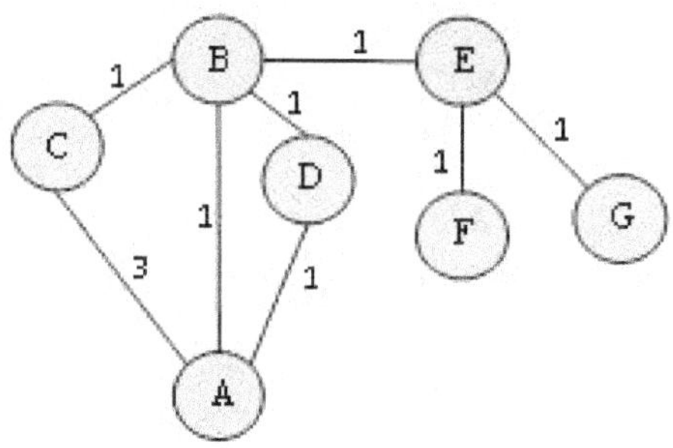

Figure 5.2: Undirected Weighted Graph

Initially, an adjacency matrix A and degree matrix D for the input graph are created. The symmetric matrix M is obtained by combining an adjacency matrix A and Degree matrix D as shown below

$$A = \begin{bmatrix} 0 & 1 & 3 & 1 & 0 & 0 & 0 \\ 1 & 0 & 1 & 1 & 1 & 0 & 0 \\ 3 & 1 & 0 & 0 & 0 & 0 & 0 \\ 1 & 1 & 0 & 0 & 0 & 0 & 0 \\ 0 & 1 & 0 & 0 & 0 & 1 & 1 \\ 0 & 0 & 0 & 0 & 1 & 0 & 0 \\ 0 & 0 & 0 & 0 & 1 & 0 & 0 \end{bmatrix} \quad D = \begin{bmatrix} 5 & 0 & 0 & 0 & 0 & 0 & 0 \\ 0 & 4 & 0 & 0 & 0 & 0 & 0 \\ 0 & 0 & 4 & 0 & 0 & 0 & 0 \\ 0 & 0 & 0 & 2 & 0 & 0 & 0 \\ 0 & 0 & 0 & 0 & 3 & 0 & 0 \\ 0 & 0 & 0 & 0 & 0 & 1 & 0 \\ 0 & 0 & 0 & 0 & 0 & 0 & 1 \end{bmatrix}$$

$$A+D = M = \begin{bmatrix} 5 & 1 & 3 & 1 & 0 & 0 & 0 \\ 1 & 4 & 1 & 1 & 1 & 0 & 0 \\ 3 & 1 & 4 & 0 & 0 & 0 & 0 \\ 1 & 1 & 0 & 2 & 0 & 0 & 0 \\ 0 & 1 & 0 & 0 & 3 & 1 & 1 \\ 0 & 0 & 0 & 0 & 1 & 1 & 0 \\ 0 & 0 & 0 & 0 & 1 & 0 & 1 \end{bmatrix}$$

The obtained matrix M is n× n symmetric matrix, which is normalized within the new range between [0, 1] by using min-max normalization technique. The min-max normalization computation is done by using equation 7.

From the matrix M, first column elements are [5, 1, 3, 1, 0, 0, 0], where the maximum value is 5 and minimum value is 0. The value of M (1, 1) i.e. 5 is normalized to ((5-0) / (5-0)) * (1-0) + 0 = 1 and the value of M (2, 1) i.e. 1 is normalized to ((1-0) / (5-0)) * (1-0) + 0 = 0.20.

Similarly computing for entire matrix, normalized matrix M_N is

$$\begin{bmatrix} 1.00 & 0.25 & 0.75 & 050 & 0.00 & 0.00 & 0.00 \\ 0.20 & 1.00 & 0.25 & 0.50 & 0.33 & 0.00 & 0.00 \\ 0.60 & 0.25 & 1.00 & 0.00 & 0.00 & 0.00 & 0.00 \\ 0.20 & 0.25 & 0.00 & 1.00 & 0.00 & 0.00 & 0.00 \\ 0.00 & 0.25 & 0.00 & 0.00 & 1.00 & 1.00 & 1.00 \\ 0.00 & 0.00 & 0.00 & 0.00 & 0.33 & 1.00 & 0.00 \\ 0.00 & 0.00 & 0.00 & 0.00 & 0.33 & 0.00 & 1.00 \end{bmatrix}$$

On normalized matrix, unfold and boosting operations are applied. The unfold operation is a simple matrix squaring (i.e. $M_N \times M_N$). Thus the unfold matrix M_U is

$$\begin{bmatrix} 1.60 & 0.81 & 1.56 & 1.13 & 0.08 & 0.00 & 0.00 \\ 0.65 & 1.32 & 0.65 & 1.10 & 0.67 & 0.33 & 0.33 \\ 1.25 & 0.65 & 1.51 & 0.43 & 0.08 & 0.00 & 0.00 \\ 0.45 & 0.55 & 0.21 & 1.23 & 0.08 & 0.00 & 0.00 \\ 0.05 & 0.50 & 0.06 & 0.13 & 1.75 & 2.00 & 2.00 \\ 0.00 & 0.08 & 0.00 & 0.00 & 0.67 & 1.33 & 0.33 \\ 0.00 & 0.08 & 0.00 & 0.00 & 0.67 & 0.33 & 1.33 \end{bmatrix}$$

The boosting operation is the matrix element squaring, illustrating by taking the square of the first column element of M_U that results with the values (1.60, 0.65, 1.25,

0.45, 0.05, 0.00, and 0.00). Squaring is done for the remaining columns of matrix M_U resulting in a matrix as shown below.

$$\begin{bmatrix} 2.56 & 0.66 & 2.44 & 1.27 & 0.01 & 0.00 & 0.00 \\ 0.42 & 1.74 & 0.42 & 1.21 & 0.44 & 0.11 & 0.11 \\ 1.56 & 0.42 & 2.29 & 0.18 & 0.01 & 0.00 & 0.00 \\ 0.20 & 0.30 & 0.05 & 1.50 & 0.01 & 0.00 & 0.00 \\ 0.00 & 0.25 & 0.00 & 0.02 & 3.06 & 4.00 & 4.00 \\ 0.00 & 0.01 & 0.00 & 0.00 & 0.44 & 1.78 & 0.11 \\ 0.00 & 0.01 & 0.00 & 0.00 & 0.44 & 0.11 & 1.78 \end{bmatrix}$$

Boosting matrix M_B is obtained by using again the same min-max normalization method. Thus the resultant Boosting Matrix M_B is

$$\begin{bmatrix} 1.00 & 0.38 & 1.00 & 0.84 & 0.00 & 0.00 & 0.00 \\ 0.16 & 1.00 & 0.17 & 0.81 & 0.14 & 0.03 & 0.03 \\ 0.61 & 0.24 & 0.94 & 0.12 & 0.00 & 0.00 & 0.00 \\ 0.08 & 0.17 & 0.02 & 1.00 & 0.00 & 0.00 & 0.00 \\ 0.00 & 0.14 & 0.00 & 0.01 & 1.00 & 1.00 & 1.00 \\ 0.00 & 0.00 & 0.00 & 0.00 & 0.14 & 0.44 & 0.03 \\ 0.00 & 0.00 & 0.00 & 0.00 & 0.14 & 0.03 & 0.44 \end{bmatrix}$$

The unfolding and boosting operations are repeated alternatively on matrix M_B until the matrix gets a stable state. Finally a stable matrix M_S is obtained as

$$\begin{array}{c} \\ \\ \\ M_S = \\ \\ \\ \\ \end{array} \begin{array}{c} \text{A B C D E F G} \\ \begin{bmatrix} 1 & 1 & 1 & 1 & 0 & 0 & 0 \\ 0 & 0 & 0 & 0 & 0 & 0 & 0 \\ 0 & 0 & 0 & 0 & 0 & 0 & 0 \\ 0 & 0 & 0 & 0 & 0 & 0 & 0 \\ 0 & 0 & 0 & 0 & 1 & 1 & 1 \\ 0 & 0 & 0 & 0 & 0 & 0 & 0 \\ 0 & 0 & 0 & 0 & 0 & 0 & 0 \end{bmatrix} \end{array}$$

From the matrix M_S the partitions are interpreted by finding the number of ones in a row and combining the corresponding indices into one group. Vertices A, B, C, and D are considered as one group or one partition since all the ones in the first row of M_S

correspond to the indices A, B, C, and D. Similarly vertices E, F, and G belong to another group forming partition 2.

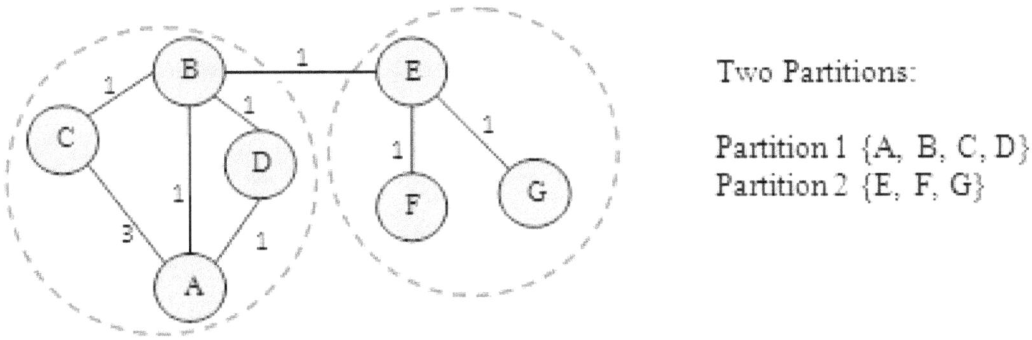

Figure 5.3: Interpreted Graph Partitions from Stable Matrix

5.5 CONCLUSION

In this chapter, a new graph partitioning approach is suggested namely Unfolding and Boosting- based Graph Partitioning algorithm (UB-GPA). The main objective is to implement UB-GPA algorithm to provide the best partitions for an object-oriented system. This can be achieved through two main operations, one is simple matrix multiplication is called unfolding and another one is squaring the elements of the unfold matrix and normalizing the element squared matrix to the range [0, 1]. These operations are repeated alternatively to achieve a stable matrix state to form clusters. The suggested UB-GPA algorithm was examined on three different software model graphs. The stable matrix occurs with low iterations while comparing with MCL algorithm iterations and it is observed that the suggested UB-GPA algorithm does not produce any overlapping groups/partitions. The results obtained were compared with existing spectral partitioning algorithm and are found to be same which can be seen in chapter VII.

The outcome of the UB-GPA algorithm can have multiple uses. 1) The partitions that are found can be utilized as reverse engineering process of the software system. 2) The partitions can be used for identification of design patterns. 3) The identified design patterns are transferred easily and utilized in another system (reusability). 4) The partitions can show some design structures like common structures, copy past structures and circular dependency structures.

6. PHASE III: DISCOVERING FREQUENT IDENTICAL DESIGN STRUCTURES

6.1 INTRODUCTION

Frequent graph mining hoist great interest in the data mining area recently. Identifying structural patterns is crucial for scientific and enterprise applications. In larger datasets it is tedious to discover the structural patterns and structural patterns are composed of subsets, patterns and trees towards networks, graphs and other problematical structures. Frequent subgraphs also known as of Frequent Itemsets.

Frequent graph pattern mining poses similar problems which are exist in recurrent item sets and mining in sequence. Based on the Apriori algorithm, a frequent n-edge label graph may contain 2^n frequent subgraphs. From the 2^n frequent subgraphs, it is difficult to identify closed frequent subgraphs. In the context of itemsets and sequence mining, an interesting solution suggested by Pasquier et.al, (1999) for mining closed frequent itemsets and sequences (Yan et al., 2003, Zaki and Hsiao, 2002, Burdick et al., 2001, Pie et al., 2000 ;). A pattern 'P' is known as closed frequent sequence if there exist no proper super-pattern of P with the redundant pattern in the dataset. The alternative approach in graph mining for finding out closed frequent subgraphs, may lead to a promising research directions as well. For example, consider a large dataset having one million frequent graphs, among million repeated graphs only 2100 are closed recurrent graphs. To analyze the large datasets, it is recommended to consider only closed frequent graphs, this will improve overall efficiency and effectiveness in finding out hidden patterns in the larger datasets.

6.2 PROPOSED APPROACH

To identify frequent closed subgraphs in large datasets to improve efficiency, better accuracy, and effectiveness, a methodology is proposed for discovering frequent closed graph structures. The proposed methodology is based on Frequent Pattern (FP) growth approach, shown in figure 6.1. FP-growth aims at isomorphic graph problems and labeled frequent subgraph mining. Proposed methodology involves generating of Depth First Search (DFS) subscription based on the partition subgraphs. Based on the subscriptions, lexicographic ordering of closed subgraphs is identified. The closed subgraphs are analyzed for finding out any right most extensions nodes. If there are any right most extensions a node exists, these nodes are identified as new closed subgraphs. Otherwise, if there is no right most extension node exists then it is considered as a redundant frequent subgraph.

6.2.1 Sequence Subgraph Subscripting

When performing a DFS on a graph, a consequent DFS tree can be constructed. When constructing a DFS tree, a linear order form of vertices is discovered by the depth-first. Proposed methodology uses the degree of the subscription to show the linear order, the discovered linear order of the vertices are assessed based on the condition of $i<j$ i.e., v_i is identified before v_j. The graph 'G' is subscripted with a DFS tree (T) which is denoted by $G_{T,}$ G_T is considered as DFS subscripting of graph G.

Multiple DFS subscripting may exists for the same graph G. Among the multiple DFS subscripting, randomly choose one of the DFS subscripting as the base subscripting. Analyze and identify the precise-most extension nodes on the base of DFS subscripting identified in the previous phase. If there is multiple right most extension nodes exist, it is tedious to reduce and construct the duplicate subgraphs. Since there are many different subscripting $G_{T,}$ that can be extended from the same graph G.

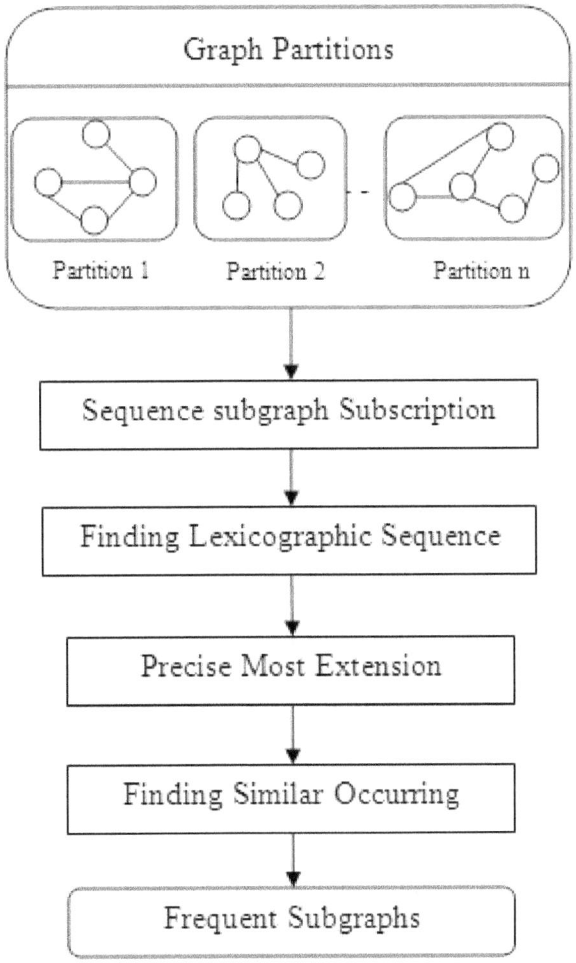

Figure 6.1: Block Diagram for Closed Subgraph Algorithm

6.2.2 Precise Most Extension

The algorithm needs extending 'G' in any possible positions, which will result in a vast number of redundant graphs. From the Graph 'g', constructed DFS tree T, it is utmost important to identify the rightmost extensions on the graph 'g'. The precise most extension also is known as a Rightmost extension. A rightmost extension for graph 'g' has two different type of extensions, these extensions are known as Forward and Backward extensions. A Backward extension is identified from the graph 'g', if there exists a node 'n' which is having an edge 'e' and also if there exist any nodes 'n_t' prior to the node 'n' from the discovered chronological order. A Forward extension is identified from the graph ‘g’, if there exists a node ‘n’ which is having an edge ‘e' and also if there exist any nodes ‘n_t’ further to the node ‘n’ from the discovered chronological order.

6.2.3 Lexicographic Sequence

The lexicographic sequence is a recursive approach to identify the sequence and ordering the subgraphs in comparison with the randomly chosen base DFS subgraph G_T as defined in section 6.2.1.

Each sequence subscripted graph can be mapped into an ordered sequence of edges. The sequences of orders are constructed and the sequence is selected that produces minimum edges compared to the base sequence. There are two kinds of order in this process: 1) Ordered Edge, that maps edges in a subscripted graph into a sequential order, and 2) Ordered Sequence, which constructs the order between sequences.

An order is constructed among the DFS sequence generated for a graph ‘g’, so that minimum DFS sequence for this graph can be defined. The proposed work is to identify accurate and efficient frequent subgraphs and to identify frequent subgraph requires finding labeled subgraphs. The label information is considered for the ordering of the weighted edges, which can be used to identify similar sequence with multiple edges, with different labels. From the proposed DFS sequence graph G_T, identified vertex l_i, l_j and edge label $l_{(i,j)}$. Consider G_T topmost precedence and the vertex label l_i as the next immediate precedence, further the edge label $l_{(i,j)}$as a third precedence and the vertex label l_j as the fourth priority to decide the regulate of two edges $l_{(i,j)}$.

6.2.4 Similar Occurrence

From the chronological order DFS Graph G_T, it is vital to find the equivalent occurrence of the subgraphs in G_T. In the closed subgraph, certain conditions are required to be satisfied for finding equivalent subgraphs if there exists any. If there are any descendent supergraphs that are not closed, these descendent supergraphs need not be verified. If the condition holds valid, then only extend graph G_{T1} is required, instead of G_T. In such scenarios it is called timely termination, since searching the supergraphs of G_T is terminated expect G_{T1}. When they are equal, it can be concluded that when G_{T1} occurs in G_T.G_{T1} must also occur exactly in the same location. Relocation of edge e in sequence G_T must be accomplished if there exists G_{T1} and should be resulted in the same sequence subgraph G_T. Therefore, some subgraph G_{T1} of G_T without the edge will not be closed since the support (G_{T1}) is not equal to support (G_T (e)).

6.3 ALGORITHM FOR CLOSED SUBGRAPH MINING

Input: A graph dataset G_D, DFS sequence D_s, its parent graph G_p, and minimum support as min_{sup}

Output: the closed frequent graph set G_S.

Method:

Step 1: Remove irregular vertices and edges

Step 2: Code of frequent graphs with single vertex and assign to S1

Step 3: for each code s ϵ S1 do

Step 3.1: if s ≠ min(s), then return

Step 3.2: if there exist e', g1=g_p (e') and g1<g_s and I (g_p, D) =L(Gp,g1,D) and g_p is not a failure case of earlier termination then return

Step 4: Set C to null

Step 5: Scan D once, find every edge e such that s can be extended to frequent s (e)l insert s(e) int C;

Step 6: Detect any possible failure of every termination in s;

Step 7: if there does not exist s(e) ϵ C, and support(s)=support(s(e)) then insert s into S;

Step 8: Remove s (e) from C which cannot be right-most extended from s;

Step 9: sort C in DFS lexicological order

Step 10: for each s(e) in C go to step 3.1

Step 11: return close graph

6.4 ILLUSTRATION OF DISCOVERING FREQUENT IDENTICAL DESIGN STRUCTURES

The proposed algorithm is illustrated with an example to discover the frequent design structures using closed graph mining technique. Let us consider Figure 6.2 with a sample labeled graph dataset, D, where two label graphs are presented.

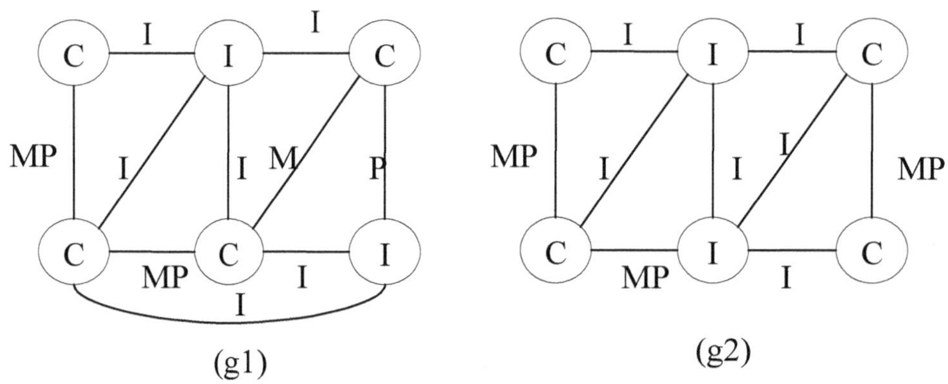

Figure 6.2: A Sample Graph Dataset D

To find Graph Subscripting:

For the performing a depth first search in a graph, construct a corresponding DFS tree. Figure 6.3(a) is repeated subgraph identified for the example graph dataset D as described in figure 6.2. The node traverse from v_0 to v_m is named as the precise most extension path. In figure 6.3 (b)-(d), three dissimilar sequences are created. The precise most path is identified by proposed methodology, from figure 6.3 (b) – (d) the rightmost paths are (v_0,v_1,v_3).

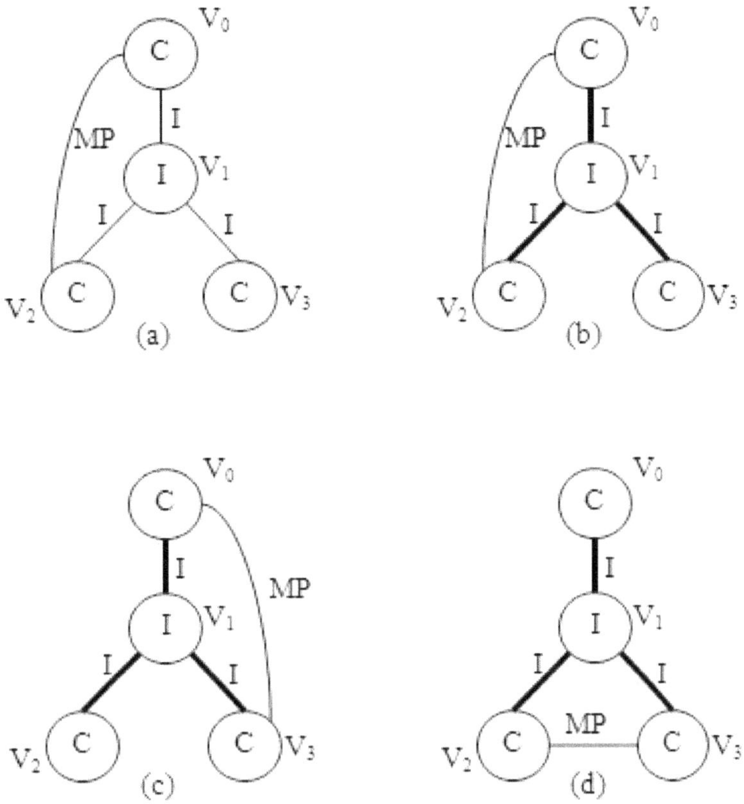

Figure 6.3: DFS Subscripts of Graph

Precise Most Extension:

The forward edge of v_i means that may present a future edge (i, j) and $i < j$ in the sequence order, the backward edge of v_i means that there exist a backward edge (i, j) and $i > j$. In figure 6.3(b), vertices (1, 3) is the forward edge of v_1, but not for v_3. To extend the figure 6.3(b), the backward extension applicant can be (v_3, v_0). The forward extension candidates can be edges expanding from v_3, v_1 or v_0 with a fresh vertex.

To find Graph Lexicographical Order:

For simplicity, an edge is represented by 5 tuples (i, j, l_i, $l_{(i,j)}$, l_j), where l_i and l_j are the labels of v_i and v_j respectively and l(i, j) is the label of the edge between them, below figure 6.4 shows the DFS subscripting for the considered graphs.

In figure 6.4, the first edges for the three DFS code are (0, 1, C, I, I), (0, 1, C, I, I) and (0, 1, C, I, I) respectively. (0, 1) the subscript is shared by all the vertices. In figure 6.4, rectangle boxes contain some values which refer to a lexicographical order.

Lexicographic order

edges	C_b	C_c	C_d
e1	(0,1, C, I, I)	(0,1,C, I, I)	(0,1,C, I, I)
e2	(1,2, C, I, I)	(1,2,C, I, I)	(1,2,C, I, I)
e3	(1,3, I, I, C)	(1,3,C, I, I)	(1,3,C, I, I)
e4	(2, 1,C, MP, C)	(3,1,C, MP, C)	(2,3,C, MP, C)

Figure 6.4 : DFS Code for Figure 6.3

To find Equivalent Occurrence:

Let g and D represent graph and dataset in figure 6.5. Recursively for g in every graph of D, the phenomenon of g in D is the sum of the number of subgraph isomorphism and is denoted by I(g, D). In figure 6.5 the graph g is occurring in dataset D. In D there are two graph g1 and g2, where g occurs in g1 one time and g occurs in g2 is two times. The total sum of the occurrences of g in D is I(g, D)=3.

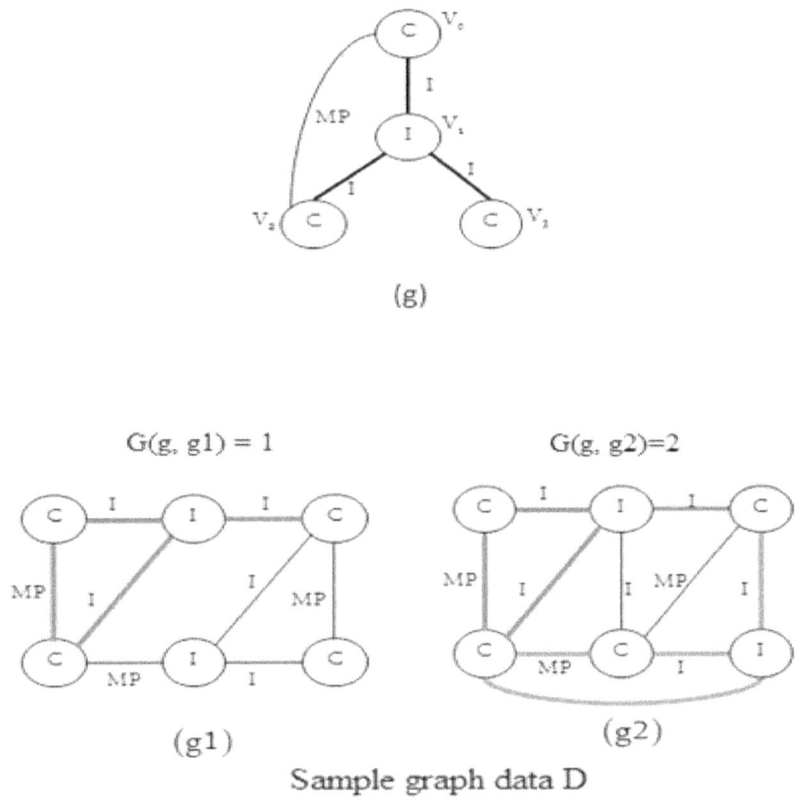

Figure 6.5 Subgraph isomorphism of g

Let G (g, g') is defined as number of possible subgraph isomorphism of g in g' Given g' = g extendable then sum of the number of extendable subgraph isomorphism, denoted by L (g, g', D). Figure 6.6 shows the two subgraphs g1 and g2, find the occurrence of g1 and g2 in sample data figure 6.6

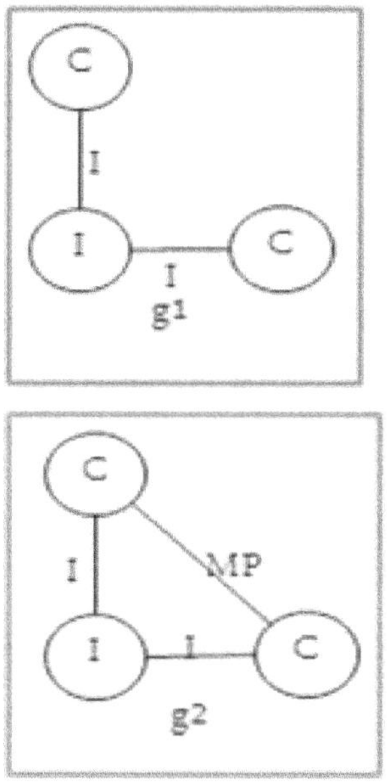

Figure 6.6: Extendable Subgraph Isomorphism

The isomorphic graphs of I (g1, D) = 2 + 2= 4, g1 is occurring in D, D contains two graphs in each graph g2 is occurring twice (2), so the total sum will be 4. Both the graphs g1 and g2 occurring in D is also 4, these graphs are an equivalent occurrence. The early termination does not hold every subgraph graph of g. Equivalent Occurrence, which means wherever g1 occurs in D, g2 occurs in the same location.

6.5 CONCLUSION

In this chapter, the proposed system is demonstrated for finding closed subgraph frequent mining patterns in large graph datasets. It is inefficient and inaccurate to mine all the frequent, equivalent, redundant structural patterns in graph mining, this is a critical problem to overcome in graph mining. The major goal of the proposed approach is to find the closed related subgraphs in a large graph data set. The algorithm includes several concepts rightmost extension and DFS lexicographical order used for frequent graph generation, equivalent occurrence and termination for non-closed graph pruning. A closed subgraph mining algorithm is tested with both hypothetical and real-world application results are discussed in Chapter VII.

7. RESULTS AND DISCUSSIONS

The possibility and effectiveness of the methodology proposed for the discovering frequent identical design structures were tested on both hypothetical and real data. The Gifviewer application is used for analyzing the performance of the proposed system.

All the experiments are done on a 2.30 GHz, Intel core i3 Personal Computer with 4 GB main memory, using Windows 7 professional. All the algorithms are developed in C# (Microsoft Visual Studio 2010) with supporting Boost library files and NRefactory.

7.1 EXPERIMENTAL STUDY OF ST-COG APPROACH

The results of experimental study applied to hypothetical and real-world data sets are discussed in this selection.

7.1.1 The analysis using Hypothetical Data

The use of the suggested ST-COG approach is discussed in detail using hypothetical data. The construction of class-oriented model graph is done for hypothetical data such as (software design 1, software design 2 and software design 3). Table 7.1 gives the information about the total number of classes (nodes) and links/ associations (edges) in the model graph.

Table 7.1: Information about Nodes and Edges of Hypothetical Data

Hypothetical Data	Number of Vertices	Number of Edges
Software Design 1	6	10
Software Design 2	9	9
Software Design 3	12	12

Software Design 1: Abstract Syntax Tree, class name/labels with their adjacency nodes

The software design is loaded into the parser, which scans entire source program and produces syntax tree based on the class name and its member functions and member variables. For Software design 1, the syntax tree is shown in Figure7.1 and similarly for the software design 2 and 3 also.

Figure 7.1: Syntax Tree for the Software Model 1

For the below considered statement, Semantic tree as follows

TypeMember: PropertyDeclartaion (Entity Type = Property. Modifiers = Public. Name = LastName)

Resolver () =MemberResolverResult

|-TatgetResult = null

|-Member = [Property ProxyPattern.Entities.Customer.LastName]

|-IsVirtualCall = False

|-Type = System.String

Table 7.2: Class Labels of Software Model 1

Class numbers	Class Labels/Names
1	OrderRepositoryBase
2	RealOrderRepository
3	ProxyOrderRepository
4	Program
5	OrderDetail
6	Order
7	Customer

Table 7.3: Vertex Numbers and Adjacent Vertices of Software Model 1

Vertex Numbers	Adjacent Nodes/Vertices
1	6, 7, 2, 3
2	1, 6, 7, 3
3	1, 2, 5, 6, 7, 4
4	3, 6
5	5, 3
6	6, 1, 2, 3. 4
7	7, 1, 2, 3

The ST-COG graph:

This class-oriented model graph gives the detail information about the class names and their type and their necessity of linking between the entities with their labels, which helps the programmers to understand about a particular application.

From the Figure 7.2 the total number of relations is shown in Table 7.4

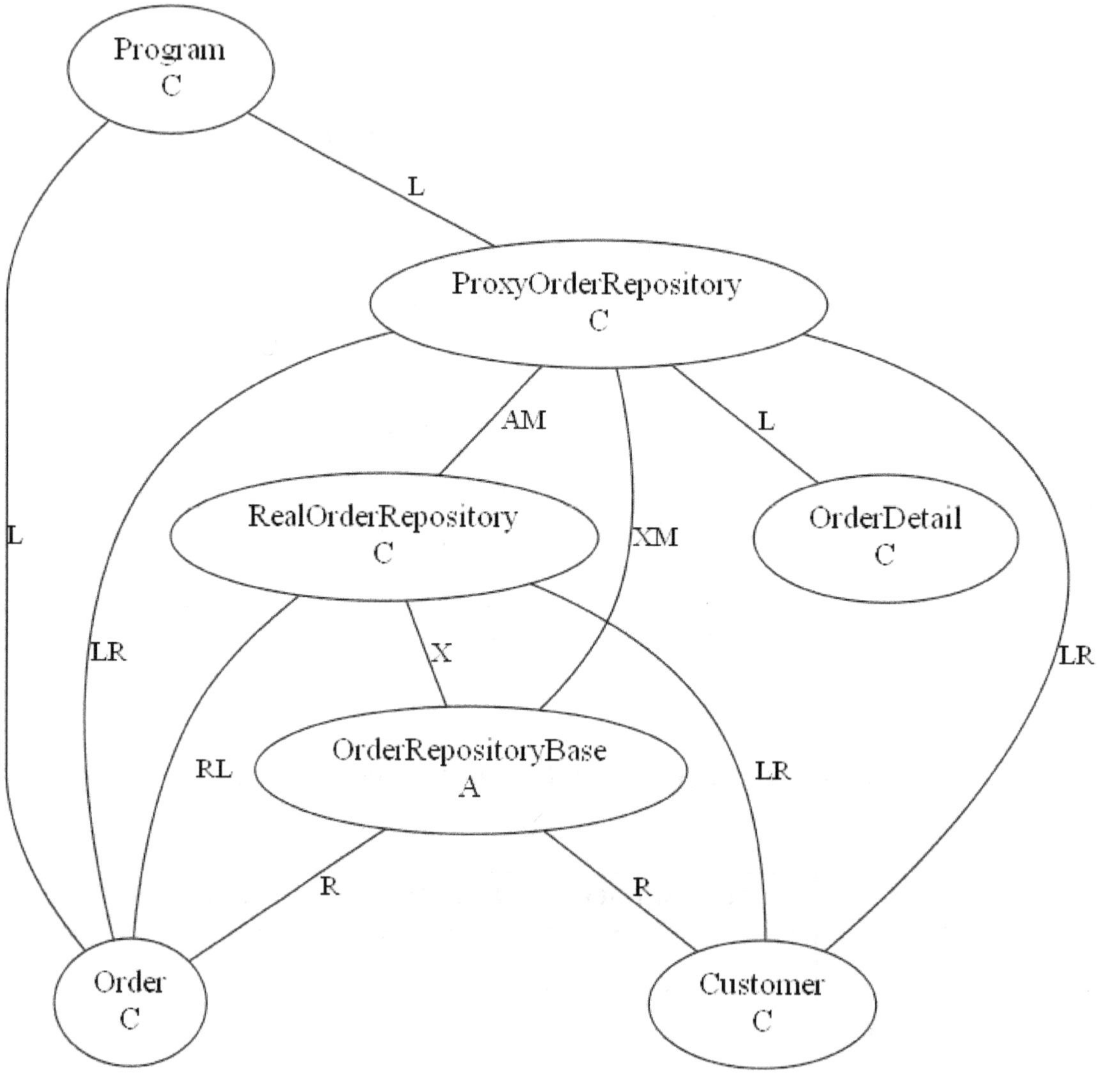

Figure 7.2: Class-Oriented Model Graph for Software Model 1

Table 7.4: Types of Relations in Software Model 1

Type of association	Software Applications		
	Total number of connections in model graph 1	**Total number of connections in model graph 2**	**Total number of connections in model graph 3**
X	2	1	2
I	0	0	0
A	1	4	5
T	0	0	0
L	5	2	1
P	0	0	0
R	6	1	1
M	2	4	7
O	0	0	3

7.1.2 The Results on Real-World Data

A class-oriented model graph is constructed for a real-world software application Gifviewer. The Gifviewer consists of a total number of nodes 47 and the total number of edges 165. The program source code is loaded into the parser to display the syntax tree and Semantic tree for the Gifviewer. The total number of class labels is shown in Table 7.5

Table 7.5: Total Number of Class Names in Gifviewer

S. no	Name of the class
1	Settings
2	Resources
3	ErrorBox
4	ErrorBox
5	FastBitmapExtensions
6	FastBitmap
7	Update
8	TimelineFrameDisplayType
9	TimelineBehaviorType
10	Knob
11	FrameChangedEventArgs
12	RangeChangeedEventArgs
13	Timelinecontrol
14	SettingsForm
15	SettingsForm
16	StreamUtils
17	Settings
18	WindowsApplicationBase
19	Program
20	GifFile
21	PackedFields
22	GifComponentStatus
23	GifDecoder
24	ErrorState
25	DisposalMethod
26	ColourTablestrategy

27	TableBasedImageData
28	NetscapeExtension
29	LogicalScreenDescriptor
30	ImageDesciptor
31	GraphicContrilExtension
32	GifHeader
33	GifFrame
34	GifComponent
35	DataBlock
36	ColourTable
37	ApplicationExtension
38	FrameExtractorHelp
39	FrameExtractorHelp
40	FrameExtract
41	FrameExtract
42	FormateForm
43	FormateForm
44	FormMain
45	FormMain
46	FileAssociation
47	CPictureBox

Gifviewer Node numbers and their adjacent nodes

Gifviewer associations between the nodes can be viewed in Table 7.6

Table 7.6: Adjacency Nodes for Gifviewer

S. no	Name of the class
1	6, 15, 20, 45
2	23
3	6,14, 20, 34, 38, 40, 42, 44, 4, 33, 41, 45
4	3, 41, 43, 45
5	6
6	1, 5, 17, 3, 14, 20, 33, 34, 38, 40, 41, 42, 44, 45, 47
7	10, 45
8	8, 13
9	9, 13
10	13, 7, 45
11	11
12	12
13	8, 9, 10, 41, 40, 44
14	3, 6, 20, 34, 40, 42, 44, 33, 41, 45
15	1, 17, 45
16	34
17	17, 6, 15, 20, 45
18	18
19	19, 46
20	1, 3, 6, 14, 17, 23, 33, 38, 40, 42, 44, 41, 45
21	21, 31
22	34, 43
23	29, 31, 32, 33, 34, 36, 20
24	24, 22, 34
25	25, 33
26	6

27	33, 34, 35
28	34, 35, 37, 23
29	34, 23
30	34, 23
31	21, 34, 23, 33
32	34, 23, 33
33	3, 6, 14, 20, 25, 30, 31, 32, 34, 36, 38, 40, 42, 44, 23, 27
34	3, 6, 14, 20, 24, 33, 35, 38, 40, 42, 44, 16, 23, 27, 28, 29, 30, 31, 32, 36, 37, 41, 43, 45
35	34, 27, 28, 37
36	34, 23, 33, 45
37	34, 35, 28
38	3,6,14,20,34,40,42,33,41,45
39	38, 41
40	3,6,13,14,20,34,38,42,44,47,33,41,45
41	3,4,6,13,14,20,34,38,39,40,42,33,44,45
42	3,6,14,20,34,38,40,44,33,41,45
43	4,22,34,41
44	3,6,13,14,20,34,40,42,47,33,41,45
45	1,3,4,6,7,10,14,15,14,20,34,36,38,40,41,42,44,46
46	19,45
47	6,40,44

For the Table 7.5 and Table 7.6 the below Figure 7.3 is obtained.

From the Figure7.3 getting the types of relation and total number of relations in the model graph can be viewed in Table 7.7

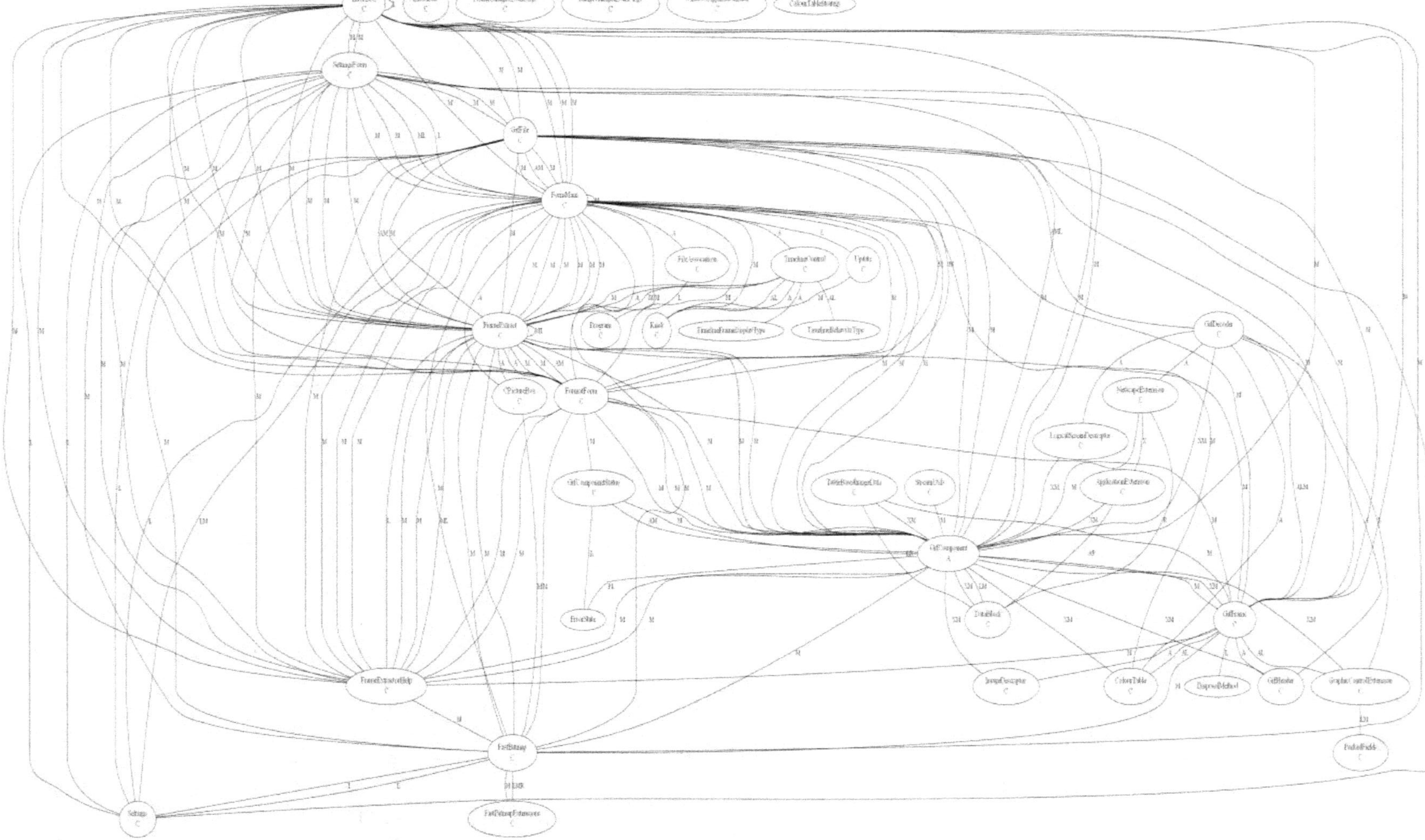

Figure 7.3: Class-Oriented Model Graph for Gifviewer

Table 7.7: Total Number of connections Obtained In the Class-Oriented Model Graph

Type of associations	Total number of connections in model graph
X	2
I	0
A	29
T	0
L	38
P	3
R	3
M	176
O	33

7.2 EXPERIMENTAL STUDY OF THE PROPOSED GRAPH PARTITIONING UB-GPA ALGORITHM

The proposed graph partitioning algorithm named UB-GPA was tested on the hypothetical data and on the real-world application and the results obtained are discussed in this section.

7.2.1 The Results on Hypothetical Data

For the obtained class-oriented model graphs of hypothetical software applications, we are applying an unfolding and boosting graph partitioning approach to partitioning the class-oriented model graph. The proposed approach Unfolding and Boosting operation based Graph Partitioning Algorithm (UB-GPA) is tested on hypothetical software application and compared with the MCL algorithm and Spectral graph partitioning algorithm. The results obtained by UB-GPA Algorithm is in Table 7.8

and UB-GPA results obtained are compared with existed algorithms are shown in Table 7.9

Table 7.8: Partitions of UB-GPA Algorithm

	No of partitions	Set of nodes (UB-GPA)
Software Design 1	Partition 1	{1, 2, 3, 4, 5, 6, 7}
Software Design 2	Partition 1	{1, 2, 3}
	Partition 2	{4, 5, 6}
	Partition 3	{7, 8, 9}
Software Design 3	Partition 1	{1, 2, 3}
	Partition 2	{ 4, 5, 7, 8}
	Partition 3	{6, 10, 11, 12}

Table 7.9: Comparison of UB-GPA with Existing Algorithm

	No of partitions	**Set of nodes (UB-GPA)**	**Set of nodes (MCL Algorithm)**	**Set of nodes (Spectral graph partitioning)**
Software Design 1	Partition 1	{1, 2, 3, 4, 5, 6, 7}	{1, 2, 3, 4, 5, 6, 7}	{1, 2, 3, 4, 5, 6, 7}
Software Design 2	Partition 1	{1, 2, 3}	{1, 2, 3}	{1, 2, 3}
	Partition 2	{4, 5, 6}	{4, 5, 6}	{4, 5, 6}
	Partition 3	{7, 8, 9}	{7, 8, 9}	{7, 8, 9}
Software Design 3	Partition 1	{1, 2, 3}	{1, 2, 3,**6**}	{1, 2, 3}
	Partition 2	{ 4, 5, 7, 8}	{ 4, 5, 7, 8}	{ 4, 5, 7, 8}
	Partition 3	{6, 10, 11, 12}	{**6**, 10, 11, 12}	{6, 10, 11, 12}

From the above Table 7.5 it is observed application 3 contain overlapping of vertex 6 in partition 1 and partition 3, with the MCL algorithm. There is no overlapping in the proposed algorithm and existed spectral graph partitioning algorithm. The proposed algorithm is compared with spectral it is observed that the partitions are same. The over summary can be view in Table 7.10.

Table 7.10: Summary Table

Hypothetical	**UB-GPA Algorithm**		**MCL Algorithm**		**Spectral Partitioning Algorithm**	
	Total number of partitions	**Occurrence of Overlapping**	**Total number of partitions**	**Occurrence of Overlapping**	**Total number of partitions**	**Occurrence of Overlapping**
Software Design 1	1	No	1	No	1	No
Software Design 2	3	No	3	No	3	No
Software Design 3	3	No	3	Yes	3	No

From Table 7.10 observed that MCL algorithm have overlapping and there is no overlapping in proposed approach UB-GPA and spectral partition.

7.2.2 The Results on Real-World Data

For the obtained class-oriented model graphs of real-world software applications, Unfolding a Boosting Graph Partitioning Algorithm (UB-GPA) is applied to partitioning the software model graph. The proposed approach is tested on GIFVIEWER software application and compared with the MCL algorithm and Spectral graph partitioning algorithm.

For the obtained ST-COG is partitioned by UB-GPA algorithm and number of partitions are shown in Table 7.11

Table 7.11: Number of Partitions with UB-GPA Algorithm

Real-world application	Total number of partitions
GIFVIEWER	4

The obtained partitions of the Gifviewer is compared with the existing algorithms of MCL and Spectral partitions algorithms are shown in Table 7.12

Table 7.12: Comparison of Gifviewer with Existed Partitions

Real-world application	UB-GPA Algorithm		MCL Algorithm		Spectral Partitioning Algorithm	
	Total number of partitions	Occurrence of Overlapping	Total number of partitions	Occurrence of Overlapping	Total number of partitions	Occurrence of Overlapping
GIFVIEWER	4	No	3	Yes	4	No

7.3 EXPERIMENTAL STUDY OF CLOSED SUBGRAPH MINING APPROACH

The proposed closed subgraph mining algorithm was tested on the hypothetical data and on the real-world application and the results obtained are discussed in this section.

7.3.1 The Results on Hypothetical Data

As observed in Table 7.6 the software design 1 there is only one partition, software design 2 there are 3 partitions and software design 3 has 3 partitions. The proposed closed subgraph approach is applied to the obtained partitioned hypothetical software design to find the identical design structures. Thus the number of frequent subgraphs obtained for software design 1, software design 2 and software design 3 are shown in Table 7.13, 7.14 and 7.15 respectively.

Table 7.13: Identification of Frequent Subgraphs for Software Design 1

Subgraph/graph	Frequency			Total frequency
	g_1	g_2	g_3	
S1	1	-	-	1

Table 7.14: Identification of Frequent Subgraphs for Software Design 2

Subgraph/graph	Frequency			Total frequency
	g_1	g_2	g_3	
S1(whole of g)	1	-	-	1
S2(g_1' excluding g)	-	1	1	2
S3 (g_2' excluding g,g1,g2)	-	-	1	1

Table 7.15: Identification of Frequent Subgraphs for Software Design 3

Subgraph/graph	Frequency			Total frequency
	g_1	g_2	g_3	
S1(whole of g)	1	-	-	1
S2(g_1' excluding g)	-	1	-	1
S3 (g_2' excluding g,g1,g2)	-	-	1	1

7.3.2 The Results on Real-World Data

From Table 7.12, four partitions are obtained for Gifviewer. The obtained result of the graph partition is considered for finding identical frequent subgraphs. The identification of frequent subgraphs for the Gifviewer application is shown in Table 7.16

Table 7.16: identification of frequent subgraphs for Gifviewer

Subgraph/graph	Frequency				Total frequency
	g_1	g_2	g_3	g_4	
S1(whole of g)	1	-	1	1	3
S2(g_1' excluding g)	-	1	-	-	1
S3 (g_2' excluding g,g_1,g_2)	-	-	1	1	2
S3 (g_3' excluding g,g_1,g_2, g_4)	-	-	-	1	1

7.4 CONCLUSION

In this chapter, the result analysis is done on hypothetical data of the proposed system and the performance is tested on real-world application Gifviewer. Experiment analysis is done on proposed system with three different hypothetical software designs. The partitions for hypothetical software designs by UB-GPA as follows: one partition is obtained for software design 1, three partitions are obtained for software design 2 and three partitions are obtained for software model 3. The results obtained do not contain any overlapping. By applying the closed subgraph mining algorithm to the obtained partitions it performed well. The performance of the proposed systems is tested on Gifviewer real-world application. The performance of the proposed system on Gifviewer is satisfactory.

8. CONCLUSIONS AND FUTURE SCOPE OF WORK

8.1 CONCLUSIONS

In this chapter, an approach is proposed to generate a class-oriented model graph and to analyze the implementation using syntax tree method. In suggested approach, the executable file of C# program is taken as input to the design system. It saves the file into memory and reads one by one to save all the implementation in order to build the syntax tree. Once the creation of abstract syntax tree is completed, it is ready for analysis. The NRefactory library is utilized to generate syntax tree. Generated syntax tree can be used for finding refactoring of similar source code and for finding a patterns design. The construction of class-oriented model graph provides in-depth information about a system. These model structured graphs can assist developers in better understanding the software architecture of the object-oriented system.

A new graph partitioning approach is suggested namely Unfolding and Boosting-based Graph Partitioning algorithm (UB-GPA). The main objective is to implement UB-GPA algorithm to provide the best partitions for an object-oriented system. This can be achieved by two main operations, one is simple matrix multiplication is called unfolding and another one is squaring the elements of the unfold matrix and normalizing the element squared matrix to the range [0, 1]. These operations are repeated alternately to achieve a stable matrix state to form clusters. The suggested UB-GPA algorithm was examined on three different software model graphs. The stable matrix occurs with low iterations while comparing with MCL algorithm iterations and it is observed that the suggested UB-GPA algorithm does not produce any overlapping groups/partitions. The results obtained were compared with existing spectral partitioning algorithm and are found to be same.

A closed subgraph mining algorithm is suggested to find similar design structures in object-oriented design (i.e. Software model). The major goal of this approach is to find the closed related subgraphs in a large graph data set. This algorithm includes several concepts rightmost extension and DFS lexicographical order used for frequent graph generation, equivalent occurrence and termination for non-closed graph pruning. The result obtained by proposed system is performed well for detection of closed graphs and satisfactory.

8.2 FUTURE SCOPE OF THE RESEARCH

The future directions for researchers are

1. We can find the similar structures by considering two different open source projects and detect the identical design structures in those projects.
2. We can develop optimize algorithms for partitioning the model graph without losing the higher relationships between the entities.

REFERENCES

[1] Agrawal, R. and Srikant, R., Fast Algorithms for Mining Association Rules in Large Databases, *In Proceedings of the 20th International Conference on Very Large Data Bases,* **pp. 487-499, (1994).**

[2] Agrawal, R., Gehrke, J., Gunopulos, D. and Raghavan, P., Automatic subspace clustering of high dimensional data for data mining applications, *In Proceedings of ACM-SIGMOD'98*, **pp.94-105, (1998).**

[3] Appala Srinuvasu Muttipati and Dr. P. Padmaja, Constructing of Software Model Graph and Analyzing Object-Oriented Programming (C#) Using Abstract Syntax Tree Method, *International Journal of Computer Science and Information Technologies,* **Vol. 6, No. 4, pp. 3288-3293, 2015**.

[4] Appala Srinuvasu Muttipati and Dr. P. Padmaja, Partitioning a Class-Oriented Model Graph (Software System) by Modified-MCL Algorithm, *International Journal of Advance Research in Computer Science and Management studies,* **Vol. 3, No. 11, pp. 29-40, 2015**.

[5] ArgoUML, http//argouml.tiggis.org

[6] Arnold, R. S., An introduction to software restructuring, *Tutorial on Software Restructuring,* **(1986).**

[7] Bader, D. A., Meyerhenke, H., Sanders, P. and Wagner, D, Graph Partitioning and Graph Clustering, **10th DIMACS Implementation Challenge Workshop, February 13-14, (2012).**

[8] Bader, M., Space-Filling Curves, *Springer*, **(2013).**

[9] Barnard, S. and Simon, H., A fast multilevel implementation of recursive spectral bisection for partitioning unstructured problems, *In Proceedings of 6th SIAM Conference Parallel Processing for Scientific Computing,* **pp 711-718, (1993).**

[10] Baxter, I. D., Yahin, A., Moura, L., Sant'Anna, M. and Bier, L., Clone detection using abstract syntax trees, In proceedings of International Conference on Software Maintenance, **pp. 368-377, (1998).**

[11] Biggerstaff, T. and Richter, C., Reusability framework, assessment, and directions, IEEE in Software, **Vol. 4, No. 2, pp. 41-49, (1987).**

[12] Booch, G., Rumbaugh, J. and Jacobson, I., The Unified Modeling Language User Guide, **Pearson Education, India, (1999).**

[13] Boppana, R. B., Eigenvalues and Graph Bisection: An Average-Case Analysis, *In Proceeding of 28th symposium on foundations of computer Science (FOCS),* **pp. 280-285, (1987).**

[14] Borgelt, C. and Berhold, M. R., Mining molecular fragments: Finding relevant substructures of molecules, *In Proceedings of IEEE International Conference on Data Mining (ICDM 02),* **pp. 51-58, (2002).**

[15] Brooks, F. P., No silver Bullet-essence and accidents of software engineering, *IEEE Computer,* **pp.10-19, (1987).**

[16] Bruegge, B. and Dutoit, H. A., Object-oriented software engineering: conquering complex and changing systems, **Prentice Hall, (2000).**

[17] Bui, T. N. and Jones, C., A heuristic for reducing fill-in in sparse matrix factorization, *In proceedings of 6th SIAM Conference on Parallel Processing for Scientific Computing,* **pp. 445-452,(1993).**

[18] Burdick. D., Calimlim, M. and Gehrke, J., MAFIA:A maximal frequent itemset algorithm for transactional database, *In Proceedings of International Conference of Data Engineering,* **pp. 443-452, (2001).**

[19] Casais, E., Managing Evolution in Object Oriented Environments: An Algorithmic Approach, **Ph.D. thesis, University of Geneva, (1991).**

[20] Chatzigeorgiou, A., Tsantalis, N. and Stephanides, Application of Graph Theory to OO Software Engineering, *In Proceedings of international workshop on interdisciplinary software engineering research,* **pp.19-36, (2006).**

[21] Chen, Y., Gansner, E. and E. Koutsofios, A C++ Data Model Supporting Reachability Analysis and Dead Code Detection, *In Proceedings of Sixth European Software Engineering Conference and Fifth ACM SIGSOFT Symposium on the Foundations of Software Engineering,* **(1997).**

[22] Chen. Y., Reverse engineering. In B. Krishnamurthy, editor, Practical Reusable UNIX Software, chapter 6, pp 177-208, John Wiley & Sons, New York, 1995.

[23] Cong, J. and Smith, M., A parallel bottom-up clustering algorithm with applications to circuit partitioning in vlsi design, *In Proceedings of ACM/IEEE Design Automation Conference*, **pp 755-760, (1993).**

[24] Cormen, T., Leiserson, C., Rivest, R. and Stein, C., Introduction to Algorithm, **2nd Edition, The MIT Press, Cambridge, MA, (2001).**

[25] Czibula, I. G., and Serban, G., Hierarchical Clustering for Software Restructuring, *INFOCOMP Journal of Computer Science*, **Vol. 6, No. 4, (2007).**

[26] Dallal, J. A., Identifying refactoring opportunities in object- oriented code: A systematic literature review, *Information and software Technology*, **Vol. 58, pp 231-249, (2015).**

[27] Daniel Grunwald, Using NRefactory for analyzing C# code (http://www.codeproject.com/Articles/408663/Using-NRefactory-for analyzing-Csharp-code)

[28] Dehaspe, L. and Toivonen.H, Discovery of frequent datalog patterns, *Data Mining and Knowledge Discovery,* **Vol. 3, No. 1, pp. 7–36, (1999).**

[29] Diekmann, R., Frommer, A. and Monien, B., Efficient schemes for nearest neighbor load balancing, *Parallel Computing*, **Vol. 25, pp. 789-812, (1999).**

[30] Diekmann, R., Monien, B. and Preis, R., Using helpful sets to improve graph bisections, Technical Report TR-RF-94-008, Department of computer science, University of Paderborn, **(1994).**

[31] Donath, W. E. and Hoffman, A. J., Algorithms for Partitioning of Graphs and Computer Logic Based on Eigenvectors of Connection Matrices, *IBM Technical Disclosure Bulletin,* **Vol. 15, No. 3, pp. 938-944, (1972).**

[32] Donath, W. E. and Hoffman, A. J., Lower Bounds for the Partitioning of Graphs, *IBM Journal of Research and Development,* **Vol. 17, No. 5, pp. 420-425, (1973).**

[33] Dongen, V. S., *Graph Clustering by Flow Simulation*, **PhD Thesis, University of Utrecht, Netherlands, (2000).**

[34] Duda, R. O., Hart, P. E. and Stork, D. G., Pattern Classification, **2nd Edition John Wiley & Sons, Inc., New York, USA, (2001).**

[35] Erdemir, U., Tekin, U. and Buzluca, F., Object oriented software clustering based on community structure, *In Proceedings Software Engineering Conference of 18th Asia Pacific*, **pp.315-321, (2011).**

[36] Ester, M., Kriegel, H. P., Sander, J. and Xu, X., Density-Based Clustering in Spatial Databases: The algorithm GDBSCAN and Its Applications, *Data Mining and Knowledge Discovery,* **Vol. 2, pp.169-194, (1998).**

[37] Farhat, C. and Lesoinne, M., Automatic Partitioning of Unstructured Meshes for the Parallel Solution of Problems in Computational Mechanics, *Journal for Numerical Methods in Engineering,* **Vol. 36, No. 5, pp. 745-764, (1993).**

[38] Fiduccia, C. M. and Mattheyses, R. M., A linear time heuristic for improving network partitions, In Proceedings of IEEE Design Automation Conference, **pp.175–181, (1982).**

[39] Fiedler, M., A Property of Eigenvectors of Nonnegative Symmetric Matrices and its Application to Graph Theory, *Czechoslovak Mathematical Journal*, **Vol. 25, No. 4, pp. 619-633, (1975).**

[40] Gamma, E. Helm, R. Johnson, R. and Vlissides, J. *Design Patterns: Elements of Reusable Object-Oriented Software*, **Addison Wesley, Boston, MA, (1995).**

[41] Gandhi, P. and Bhatia, P. K. Optimization of Object- Oriented Design Using Coupling Metrics, *International Journal of Computer Applications,* **Vol. 27, No.10, pp. 41-44, (2011).**

[42] Gibson, D., Kleinberg, J. and Raghavan, P. Inferring web communities from link topology, In Proceedings of HYPERTEXT'98, **pp.225–234, (1998).**

[43] Gilbert, J. R., Miller, G. L. and Teng, S. H., Geometric Mesh Partitioning: Implementation and Experiments, SIAM Journal on Scientific Computing, **Vol. 19, No. 6, pp. 2091-2110, (1998).**

[44] Glover, F., Tabu search-part I, *ORSA Journal of computation,* **Vol. 1, pp.190-206, (1989).**

[45] Glover, F., Tabu search-part II, *ORSA Journal of computation*, **Vol. 2, pp.4-32 (1990).**

[46] Gupta, A., Fast and effective algorithms for graph partitioning and sparse matrix reordering, *IBM Journal of Research and Development*, **Vol. 41, No. 1/2, pp. 171-183, (1996).**

[47] Han, S., Wee, K. N. and Yu, Y., FSP: Frequent Substructure Pattern Mining, *In Proceedings of the 9th International Conference on Information and Communication Security,* **pp. 1-5, (2007).**

[48] Hauck, S and Borriello, G., An evaluation of bipartitioning technique, *In Proceedings of Chapel Hill Conference on Advanced Research in VLSI*, **(1995).**

[49] Hendrickson, B. and Leland, R., A multilevel algorithm for partitioning graphs, *In Proceedings of Supercomputing*, **(1995).**

[50] Hendrickson, B. and Leland, R., An Improved Spectral Graph Partitioning Algorithm for Mapping Parallel Computations, *SIAM Journal on Scientific Computing*, **Vol. 16, No. 2, pp. 452-469, (1995).**

[51] Holder, L. B., Cook, D. J. and Djoko, S., Substructure Discovery in the Subdue System, *In Proceeding of the AAAI'94 workshop knowledge discovery in databases (KDD'94),* **pp 169–180, (1994).**

[52] Hsieh, H. P. and Li, C. T., Mining Temporal Subgraph Patterns in Heterogeneous Information Networks, *In Proceedings of IEEE International Conference on Social Computing / International Conference on Privacy, Security, Risk and Trust*, **pp. 282-287, (2010).**

[53] Huan, J., Wang, G., and Prins, J., Efficient Mining of Frequent Subgraphs in the Presence of Isomorphism, *In Proceedings of the 3rd IEEE International Conference on Data Mining,* **pp. 549-552, (2003).**

[54] Huan, J., Wang, W., Prins, J. and Yang, J., Spin: Mining maximal frequent subgraphs from graph databases, UNC Technical Report TR04-018, **(2004).**

[55] Hungershofer, J. and Wierum, J. M., On the Quality of Partitions based on Space-Filling Curves, *In proceedings of International Conference on Computational Science-ICCS,* **pp. 36-45, (2002).**

[56] Inokuchi, A., Washio, T. and Motoda, H., An apriori-based algorithm for mining frequent substructures from graph data, *In Proceedings of European Conference on Principles of Data Mining and Knowledge Discovery (PKDD 00),* **pp. 13-23, (2000).**

[57] Jeffrey L. Overbey and Ralph E. Johnson, Generating Rewritable Abstract Syntax Trees, *Software Language Engineering,* Lecture Notes in Computer Science, Springer- Verlag Berlin Heidelberg, **Vol.5452, pp 114-133, (2009).**

[58] Jiawei, H. and Kamber, M., *Data mining: concepts and techniques, San Francisco, CA,* ***itd: Morgan Kaufmann*** **(2001).**

[59] Johnson, D. S., Argon, C. R., McGeoch, L. A. and Schevon, C., Optimization by simulated annealing: An experimental evaluation; Part I, graph partitioning, *Operation Research,* **Vol. 37, No. 6, pp. 865-892, (1989).**

[60] Johnson, R. E. and Foote, B., Designing reusable classes, Journal of Object-Oriented Programming, **Vol. 1, No. 2, pp. 22-35, (1988).**

[61] Karypis, G. and Kumar, K., A parallel algorithm for multilevel graph partitioning and sparse matrix ordering, *Journal of Parallel and Distributed Computing,* **Vol. 48, No. 1, (1998b).**

[62] Karypis, G. and Kumar, K., Multilevel k-way partitioning scheme for irregular graphs, *Journal of Parallel and Distributed Computing*, **Vol. 48, No. 1, (1998a).**

[63] Kernighan, B. W. and Lin, S., An efficient heuristic procedure for partitioning graphs, T*he Bell System Technical Journal,* **Vol. 49, No. 0, pp. 291–307, (1970).**

[64] Kirkpatrick, S., Gelatt, C. D. and Vecchi, M. P., Optimization by simulated annealing, *Science,* **Vol. 220, No. 4598, pp. 671-680, (1983).**

[65] Kirmani, S. and Raghavan, P., Scalable Parallel Graph Partitioning, *In Proceedings of High Performance Computing, Networking, Storage and Analysis, ACM,* **(2013).**

[66] Kleinberg, J. M. and Tardos, E., Approximation algorithms for classification problems with pairwise relationships: Metric labeling and Markov random fields, *Journal of the ACM,* **Vol. 49, No. 9, pp.14-23, (2002).**

[67] Kleppe, A. and Rensink, A., On a Graph-Based Semantics for UML Class and Object Diagrams, *In Proceedings of Seventh International Workshop on Graph Transformation and Visual Modeling Techniques,* **pp.1-16, (2008).**

[68] Korn, J., Chen, Y. and Koutsofios, E., Chava: Reverse Engineering and Tracking of Java Applets, *In Proceedings of the 6th Working Conference on Reverse Engineering*, **pp 314–325, (1999).**

[69] Kuramochi, M. and Karypis, G., Finding Frequent Patterns in a Large Sparse Graph, *In Proceedings of the4th SIAM International Conference on Data Mining (SDM 2004)*, **(2004).**

[70] Kuramochi, M. and Karypis, G., Frequent Subgraph Discovery, *In Proceedings of IEEE International Conference on Data Mining,* **pp. 313-320, (2001).**

[71] Kuramochi, M. and Karypis, G., GREW: A Scalable Frequent Subgraph Discovery Al gorithm, *In Proceeding of the international conference on data mining (ICDM'04)*, **pp.439–442, (2004).**

[72] Lam, W. W. M. and Chan, K. C. C, Discovering Interesting Molecular Substructures for Molecular Classification, *IEEE Transactions on Nanobioscience*, **Vol. 9, No. 2, pp. 77-89, (2011).**

[73] Lanczos, C An Iteration Method for the Solution of the Eigenvalue Problem of Linear Differential and Integral Operators, *Journal of Research of the National Bureau of Standards,* **Vol. 45, No. 4, pp. 255-282, (1950).**

[74] Leland, R. and Hendrickson, B., An empirical study of static load balancing algorithms, *In proceeding of scalable High-performance computing Conference*, **pp. 682-685, (1994).**

[75] Li, Y., Lin, Q., Zhong, G., Duan, D., Jin, Y and Bi, W.,A directed labeled graph frequent pattern mining algorithm based on minimum code, *In Proceedings of International Conference on Multimedia and Ubiquitous Engineering (ICMUE '09)*, **pp. 353-359, (2009).**

[76] Li, Z., Lu, S., Myagmar, s. and Y. Zhou,y., CP-Miner: Finding copy–paste and related bugs in large scale software code, *IEEE Transaction of Software Engineering*, **Vol. 32, pp. 172-192, (2006).**

[77] Liu, Y., Li, J. and Gao, H., JPMiner: Mining Frequent Jump Patterns from Graph Databases, *In Proceedings of International Conference on Fuzzy Systems and Knowledge Discovery (ICFSKD '09),* **pp. 114-118, (2009).**

[78] Luo, B., Wilson, R. C. and Hancock, E. R., Spectral clustering of graphs, in: G. Goos, J. Hartmanis, J. van Leeuwen (Eds.), Proceedings of the Tenth International Conference on Computer Analysis of Images and Patterns, CAIP, in: Lecture Notes in Computer Science, **Vol. 2756, Springer Verlag GmbH, Berlin, Heidelberg, Germany, (2003)**.

[79] Luo, B., Wilson, R. C. and Hancock, E. R., Spectral Features Vectors for Graph Clustering, in: T. Caelli, A. Amin, R.P. Duin, M. Kamel, D. de Ridder (Eds.), Proceedings of the Joint IARP International Workshops on Syntactical and Structural Pattern Recognition and Statistical Pattern Recognition, in: Lecture Notes in Computer Science, **Vol. 2396, Springer Verlag GmbH, Berlin, Heidelberg, Germany, (2002).**

[80] Miller, G., Teng, S. H. and Vavasis, S., A Unified Geometric Approach to Graph Separators, *In proceedings of 32nd Symposium on Foundations of Computer Science (FOCS),* **pp. 538-547, (1991).**

[81] Muggleton, S. and Feng, C., Efficient induction of logic programs, *In Proceedings of the Workshop on Algorithmic Learning Theory*, **pp. 368–381, (1990).**

[82] Muggleton, S., Stochastic logic programs, *In L. de Raedt, ed. Advances in Inductive Logic Programming.* IOS Press, Amsterdam, **(1996).**

[83] Neamtiu, I., Foster, J. S. and Hicks, M, Understanding Source Code Evolution Using Abstract Syntax Tree Matching, *In Proceedings of international workshop on Mining software repositories* (MSR '05). ACM, New York, NY, USA, **pp. 1-5, (2005).**

[84] Ng, R. T. and Han, J. Efficient and effective clustering method for spatial data mining, *In Proceedings of VLBD'94,* **pp.144–155, (1994).**

[85] Nijssen, S. and Kok, J., A quickstart in frequent structure mining can make a difference, *In Proceedings of ACM International Conference on Knowledge Discovery and Data Mining (SIGKDD 04)*, **pp. 647–652, (2004).**

[86] Nijssen, S. and Kok, J., Faster association rules for multiple relations, *In Proceedings of Seventeenth International Joint Conference on Artificial Intelligence*, **Vol. 2, pp. 891–896, (2001).**

[87] Parnas, D. L., Designing software for ease of extension and contraction, *IEEE Transactions on Software Engineering,* **Vol. 5, No. 2, pp. 128–138, (1979).**

[88] Pasquier, N., Bastide, Y., Taouil, R. and Lakhal, L., Discovering frequent closed itemsets for association rule, *In proceedings of ICDT*, **(1999).**

[89] Pei, J., Han, J. and Mao, R., CLOSET: An efficient algorithm for mining frequent closed itemsets, In *Proceedings of DMKD*, **pp. 21-30, (2000).**

[90] Pilkington, J. R. and Baden, S. B., Partitioning with Space-filling Curves, *Technical Report CS94-349, UC San Diego, Department of Computer Science and Engineering,* **(1994).**

[91] Pothen, A., Simon, H. and Liou, K., Partitioning sparse matrices with eigenvectors of graphs, *SIAM Journal of Matrix Analysis and Applications,* **Vol. 11, No.3, pp. 430-452, (1990).**

[92] Quinlan, J. R., Learning logical definitions from relations, *Machine Learning,* **Vol. 5, No. 3, pp. 239–266, (1990).**

[93] Rajani, B. and Roopa, M.A., Source Code to Module Dependency Graph Using Class Dependency Analyzer, *International journal of Advanced Research in Computer Science and Software Engineering,* **Vol.3, No.7, pp.197-201, (2013).**

[94] Riel, A. J., Object-Oriented Design Heuristics, **AddisonWesley, Boston, MA, (1996).**

[95] Robles-Kelly, A. and Hancock, E. R., Graph edit distance from spectral seriation, IEEE Transactions on Pattern Analysis and Machine Intelligence, **Vol. 27, No. 3, pp. 365–378, (2005).**

[96] Rolland, E., Pirkul, H. and Glover, F, Tabu search for graph partitioning, *Annals of Operations Research,* **Vol. 63, No. 2, pp. 209-232, (1996).**

[97] Saxena, V. and Kumar, S., Impact of Coupling and Cohesion in Object-Oriented Technology, *Journal of Software Engineering and Applications,* **Vol. 5, pp. 671-676, (2012).**

[98] Schaeffer, S. E., Graph Clustering, *Computer Science Review,* **pp. 27-64, (2007).**

[99] Schamberger, S. and Wierum, J. M., A Locality Preserving Graph Ordering Approach for Implicit Partitioning: Graph-Filling Curves, *In Proceedings of 17th International Conference on Parallel and Distributed Computing Systems (PDCS),* **pp. 51-57, (2004).**

[100] Schulz, C., Lowe, M. and König, H., Refactoring object-oriented systems, In: Manipulation of Graphs, Algebras and Pictures. Essays Dedicated to Hans-Jorg Kreowski on the Occasion of His 60th Birthday, Universität Bremen, **pp. 321–340, (2009).**

[101] Shi, J., Malik, J., Normalized cuts and image segmentation, *IEEE Transactions on Pattern Analysis and Machine Intelligence,* **Vol. 22, No. 8, pp. 888-905, (2000).**

[102] Simon, H. D., Partitioning of unstructured problems for parallel processing, *Computing systems in Engineering,* **Vol. 2, No. 2-3, pp. 135-148, (1991).**

[103] Singh, P., Singh, S. and Kaur, J, Tool for generating code metrics for C# source code using abstract syntax tree technique, *SIGSOFT Software Engineering* Notes 38, **Vol. 5, pp.1-6, (2013).**

[104] Thomas L, Valluri S, Karlapalem K., Isg: Itemset based subgraph mining, Technical Report, Center for Data Engineering, IIIT, Hyderabad, **(2009).**

[105] Thomas L, Valluri S, Karlapalem K., MARGIN: Maximal Frequent Subgraph Mining, *In Proceedings of Sixth International Conference on Data Mining,* **pp.1097-1101, (2006).**

[106] Virdi, H. S., and Singh, B., Study of the Different Types of Coupling Present in the Software Code, *International Journal of computer Science and Information Technology*, **Vol. 3, No. 3, pp. 4153- 4156, (2012).**

[107] Wei, Y. C., Cheng, C, K., Towards efficient hierarchical designs by ratio cut partitioning, *In Proceedings of the IEEE International Conference on Computer-Aided Design*, **pp. 298-301, (1989).**

[108] Williams, R. D., Performance of dynamic loading algorithms for unstructured mesh calculations, *Concurrency: Practice and Experience*, **Vol. 3, No. 5, pp. 457-481, (1991).**

[109] Xanthos, S., *Clustering Object-Oriented Software Systems using Spectral Graph Partitioning*, In ACM Student Research Competition, **(2005).**

[110] Yan, X. and Han, J., CloseGraph: Mining closed frequent graph patterns, *In Proceedings of ACM SIGKDD International Conference Knowledge Discovery and Data Mining (KDD 03)*, **pp. 286-295, (2003).**

[111] Yan, X. and Han, J., gSpan: Graph-based substructure pattern mining, *In Proceedings of IEEE International Conference on Data Mining (ICDM 02)*, **pp. 721-724, (2002).**

[112] Yoshida, K., Motoda, H. and Indurkhya, N., Graph-based Induction as a Unified Learning Frame work, *Journal of Applied Intelligence*, **Vol. 4, No. 3, pp.297–328, (1994).**

[113] Zaki, M. J. and Hsiao, C. J., CHARM: An efficient algorithm for closed itemset mining, *In Proceedings of SDM,* **pp.457-473, (2002).**

[114] Zhang, S., Yang, J. and Li, S., RING: An Integrated Method for Frequent Representative Subgraph Mining, *In Proceedings of International Conference on Data Mining (ICDM '09),* **pp. 1082-1087, (2009).**

[115] Zumbusch, G., Parallel Multilevel Methods: Adaptive Mesh Refinement and Load balancing, *Teubner*, **(2003).**

www.ingramcontent.com/pod-product-compliance
Lightning Source LLC
LaVergne TN
LVHW082349200926
848481LV00012B/3508